Lecture Notes in Business Information Processing   **284**

Series Editors

Wil M.P. van der Aalst
Eindhoven Technical University, Eindhoven, The Netherlands

John Mylopoulos
University of Trento, Trento, Italy

Michael Rosemann
Queensland University of Technology, Brisbane, QLD, Australia

Michael J. Shaw
University of Illinois, Urbana-Champaign, IL, USA

Clemens Szyperski
Microsoft Research, Redmond, WA, USA

W0230439

More information about this series at http://www.springer.com/series/7911

David Aveiro · Robert Pergl
Giancarlo Guizzardi · João Paulo Almeida
Rodrigo Magalhães · Hans Lekkerkerk (Eds.)

Advances in Enterprise Engineering XI

7th Enterprise Engineering Working Conference, EEWC 2017
Antwerp, Belgium, May 8–12, 2017
Proceedings

 Springer

Editors
David Aveiro
University of Madeira and Madeira
  Interactive Technologies Institute
Funchal
Portugal

Robert Pergl
Czech Technical University in Prague
Prague
Czech Republic

Giancarlo Guizzardi
Free University of Bozen-Bolzano
Bolzano
Italy

João Paulo Almeida
Federal University of Espírito Santo
Vitoria
Brazil

Rodrigo Magalhães
Kuwait Maastricht Business School
Kuwait
Kuwait

Hans Lekkerkerk
Radboud University Nijmegen
Nijmegen
The Netherlands

ISSN 1865-1348   ISSN 1865-1356   (electronic)
Lecture Notes in Business Information Processing
ISBN 978-3-319-57954-2   ISBN 978-3-319-57955-9   (eBook)
DOI 10.1007/978-3-319-57955-9

Library of Congress Control Number: 2017938323

Printed on acid-free paper

This Springer imprint is published by Springer Nature
The registered company is Springer International Publishing AG
The registered company address is: Gewerbestrasse 11, 6330 Cham, Switzerland

Preface

The CIAO! Enterprise Engineering Network (CEEN) is a community of academics and practitioners who strive to contribute to the development of the discipline of Enterprise Engineering (EE), and to apply it in practice. The aim is to develop a holistic and general systems theory-based understanding on how to (re)design and run enterprises effectively. The ambition is to develop a consistent and coherent set of theories, models, and associated methods that: enable enterprises to reflect, in a systematic way, on how to realize improvements; and assist them, in practice, in achieving their aspirations.

In doing so, sound empirical and scientific foundations should underlie all efforts and all organizational aspects that are relevant should be considered, while combining already existing knowledge from the scientific fields of information systems, software engineering, management, as well as philosophy, semiotics, and sociology, among others. In other words, the (re)design of an enterprise and the subsequent implementation of changes should be the consequence of rationalized decisions that: take into account the nature and reality of the enterprise and its environment; and respect relevant empirical and scientific principles.

Enterprises are taken to be systems whose reality has a dual nature by being simultaneously, on one hand, centrally and purposefully (re)designed, and, on the other hand, emergent in a distributed way, given the fact that, its main agents, the humans that are the pearls of the organization, act with free will, in a creative and in a responsible (or sometimes not) way. We acknowledge that, in practice, the development of enterprises is not always a purely rational/evidence-based process. As such, we believe the field of EE aims to provide evidence-based insights into the design and evolution of enterprises and the consequences of different choices irrespective of the way decisions are made.

The origin of the scientific foundations of our present body of knowledge is the CIAO! Paradigm (Communication, Information, Action, Organization) as expressed in our Enterprise Engineering Manifesto and the paper: "The Discipline of Enterprise Engineering." In this paradigm, organization is considered to emerge in human communication, through the intermediate roles of information and action. Based on the CIAO! Paradigm, several theories have been developed, and are still being proposed. They are published as technical reports.

The CEEN welcomes proposals of improvements to our current body of knowledge, as well as the inclusion of compliant and alternative views, always keeping in mind the need to maintain global systemic coherence, consistency, and scientific rigor of the entire EE body of knowledge, as a prerequisite for the consolidation of this new engineering discipline. Yearly events like the Enterprise Engineering Working Conference and associated Doctoral Consortium are organized to promote the presentation of EE research and application in practice, as well as discussions on the contents and current state of our body of theories and methods.

Since 2005 the CEEN has organized the CIAO! Workshop and, since 2008, its proceedings have been published as *Advances in Enterprise Engineering* in the Springer LNBIP series. From 2011 on, this workshop was replaced by the Enterprise Engineering Working Conference (EEWC). This volume contains the proceedings of the 7th EEWC, held in Antwerp, Belgium. There were 40 submissions. Each submission was reviewed by three Program Committee members and the decision was to accept 12 full papers and 4 short papers, which were carefully reviewed and selected for inclusion in this volume.

The EEWC aims at addressing the challenges that modern and complex enterprises are facing in a rapidly changing world. The participants of the working conference share a belief that dealing with these challenges requires rigorous and scientific solutions, focusing on the design and engineering of enterprises. The goal of EEWC is to stimulate interaction between the different stakeholders, scientists as well as practitioners, interested in making EE a reality.

May 2017

David Aveiro
Robert Pergl
Giancarlo Guizzardi
João Paulo Almeida
Rodrigo Magalhães
Hans Lekkerkerk

Organization

EEWC 2017 was the seventh Working Conference resulting from a series of successful CIAO! Workshops and EEWC Conferences over the last years. These events were aimed at addressing the challenges that modern and complex enterprises are facing in a rapidly changing world. The participants in these events share the belief that dealing with these challenges requires rigorous and scientific solutions, focusing on the design and engineering of enterprises.

This conviction has led to the effort of annually organizing an international working conference on the topic of enterprise engineering, in order to bring together all stakeholders interested in making enterprise engineering a reality. This means that not only scientists are invited, but also practitioners. Next, it also means that the conference is aimed at active participation, discussion, and exchange of ideas in order to stimulate future cooperation among the participants. This makes EEWC a working conference contributing to the further development of enterprise engineering as a mature discipline.

The organization of EEWC 2017 and the peer review of the contributions to the conference were accomplished by an outstanding international team of experts in the fields of enterprise engineering. The organizational structure of EEWC 2017 is listed herein.

Advisory Board

Antonia Albani	University of St. Gallen, Switzerland
Jan Dietz	Delft University of Technology, The Netherlands

Conference Chairs

Jan Verelst	University of Antwerp, Belgium
Henderik A. Proper	Luxembourg Institute of Science and Technology, Luxembourg

Program Chairs

David Aveiro	University of Madeira and Madeira Interactive Technologies Institute, Portugal
Robert Pergl	Czech Technical University in Prague, Czech Republic

Session Chairs

Foundational Ontologies

Giancarlo Guizzardi   Free University of Bozen-Bolzano, Italy
João Paulo Almeida   Federal University of Espírito Santo, Brazil

Organizational Design

Rodrigo Magalhes   Kuwait Maastricht Business School, Kuwait
Hans Lekkerkerk   Radboud Universiteit Nijmegen, The Netherlands

Organizing Chair

Jan Verelst   University of Antwerp, Belgium

Program Committee

Alberto Silva   INESC and University of Lisbon, Portugal
Carlos Mendes   University of Lisbon, Portugal
Christian Huemer   Vienna University of Technology, Austria
Duarte Gouveia   University of Madeira, Portugal
Eduard Babkin   Higher School of Economics, Nizhny Novgorod, Russia
Fernanda Araujo Baiao   UNIRIO, Brazil
Florian Matthes   Technical University Munich, Germany
Frank Harmsen   Maastricht University and Ernst & Young Advisory,
                            The Netherlands
Frederik Gailly   Ghent University, Belgium
Geert Poels   Ghent University, Belgium
Giancarlo Guizzardi   Free University of Bozen-Bolzano, Italy
Gil Regev   École Polytechnique Fédérale de Lausanne, Switzerland
Graham McLeod   University of Cape Town and Inspired.org, South Africa
Hans Mulder   University of Antwerp, Belgium
Jan Dietz   Delft University of Technology, The Netherlands
Jan Hoogervorst   Sogeti Netherlands, The Netherlands
Jens Gulden   University of Duisburg-Essen, Germany
Joop de Jong   Mprise, The Netherlands
Jose Tribolet   INESC and University of Lisbon, Portugal
Joseph Barjis   Institute of Engineering and Management, San Francisco,
                            USA
Julio Nardi   Federal Institute of Espírito Santo, Brazil
Junichi Iijima   Tokyo Institute of Technology, Japan
Linda Terlouw   Delft University of Technology, The Netherlands
Luiz Olavo Bonino   VU University of Amsterdam, The Netherlands
Marcela Vegetti   Universidad Tecnológica Nacional, Argentina

Martin Cloutier	Université du Québec à Montréal, Canada
Martin Op 'T Land	Capgemini, The Netherlands
Mauricio Almeida	Federal University of Minas Gerais, Brazil
Miguel Mira Da Silva	INESC and University of Lisbon, Portugal
Monika Kaczmarek	University Duisburg Essen, Germany
Nelson King	Khalifa University, United Arab Emirates
Niek Pluijmert	INQA Quality Consultants, The Netherlands
Peter Loos	University of Saarland, Germany
Petr Kremen	Czech Technical University in Prague, Czech Republic
Philip Huysmans	University of Antwerp, Belgium
Ricardo Falbo	Federal University of Espírito Santo, Brazil
Robert Lagerström	KTH Royal Institute of Technology, Sweden
Robert Pergl	Czech Technical University in Prague, Czech Republic
Robert Winter	University of St. Gallen, Switzerland
Rodrigo Magalhaes	Kuwait Maastricht Business School, Kuwait
Rony Flatscher	Wirtschaftsuniversität Wien, Austria
Sérgio Guerreiro	INESC and University of Lisbon, Portugal
Sanetake Nagayoshi	Shizuoka University, Japan
Steven van Kervel	Formetis, The Netherlands
Sybren de Kinderen	University of Luxembourg, Luxembourg
Tatiana Poletaeva	Higher School of Economics, Nizhny Novgorod, Russia
Ulrik Franke	Swedish Defense Research Agency, Sweden

Contents

Ontologies

Organisation Design

Formalisms

Formal Specification of DEMO Process Model and Its Submodel

Towards Algebra of DEMO Models

Tetsuya Suga[✉] and Junichi Iijima

Tokyo Institute of Technology, 2-12-1 Ookayama,
Meguro-ku, Tokyo 152-8550, Japan
{suga.t.ac,iijima.j.aa}@m.titech.ac.jp

Abstract. This paper discusses a specification and merge operation over submodels of a given Process Model (PM) in Design and Engineering Methodology for Organizations (DEMO). In general, a submodel is a part of a given model. An earlier work proposed how submodels of a given DEMO Construction Model (CM) can be attained by a set-theoretic formalization. However, it remains unclear how to expand the formalism to the notion of submodels of a given PM. Since the given PM should align with the corresponding CM, a submodel of the given PM should not only be a PM, but also conform to the corresponding submodel of the CM. These two independent constraints indicate the desired definition and formalization of submodels of PMs. The proposed approach is shown to be applicable to a common demonstration case. Through the formalization, this paper shows the closure, commutativity, and associativity of the merge operation over submodels of a given PM. Moreover, it is found that the consistency between CMs and PMs is preserved during the merge operation.

Keywords: Enterprise ontology · Merge · Algebra · DEMO process model · Match

1 Introduction

In recent years, organizations have been objects of interest in a variety of disciplines, not only in social science and business administration, but also in engineering. Enterprise engineering (EE) is an emerging discipline of systems engineering that studies organizations from an engineering perspective [1]. Since Enterprise Ontology provides models of enterprises with considerably reduced complexity [2], a (re)design process with such a methodology inevitably requires working with a model of the organization rather than the real-world organization itself. Put another way, (re)designing an enterprise is substantially achieved by *building and editing models of the enterprise*, and implementing the changes in the model in the real world.

© Springer International Publishing AG 2017
D. Aveiro et al. (Eds.): EEWC 2017, LNBIP 284, pp. 3–17, 2017.
DOI: 10.1007/978-3-319-57955-9_1

A vast majority of the work in this area has focused on artifacts in the category of Way of Thinking (WoT), Way of Modeling (WoM), and Way of Working (WoW) in five ways in information system methodologies [3], such as artifacts that are typically involved in the process of modeling. There is, however, relatively limited research investigating theories that may support the process of manipulating models, a possible contribution to Way of Supporting (WoS). In particular, the requirements of submodels of enterprise models and operations over the submodels remain unclear.

In this article, a **submodel** is a *part of* a given model that has been already created. The given model is often called the *global* model. Submodels are used for diverse reasons and purposes, such as to partition the enterprise, to focus on only some part in question, to hide some part for readability or even because of an access control policy, and so on. Logically speaking, regardless of the reasons, the requirements of those partial models of enterprise models should be specified if we rigidly define what "part of" means. Unless they are defined properly, we might allow submodels that do not conform to preconditions of the global model, including (but not limited to) its metamodel. An **operation** over one or more models (i.e., inputs) means creating another model by input model(s). Once the requirements of submodels have been established, we may leverage those submodels by considering operations over the submodels such as merge submodels. Such operations may include merge, diff, split, slice, and so forth. Again, if we do not define the operation rigorously, we might get a model that does not conform to the requirements of the global model as a result of performing the operation. In addition, we should study and then be aware of properties of the operation such as the associative and the commutative properties.

Insufficient investigation of the formalization and manipulability of enterprise models, i.e., a scarcity in WoS, stands in the way of the success of EE in the market. Presently, such a model manipulation and the following model checking require manual work by professionals with considerable expertise not only in the business, per se, but also in the modeling framework and theories behind it. However a certain part of the work could be taken over or even rendered unnecessary by formalization and automation with computer-aided design tools. Other disciplines, such as mechanical engineering, electrical engineering, and computer science, demonstrate the successful implementation of computer-aided design tools and automation.

Therefore, this research aims at investigating the structure and properties in submodels and operations in DEMO PMs based on an earlier work on those in DEMO CMs. Although the earlier work established the formalization of CMs, PMs are characterized by components that are specific to PMs, such as process step kinds and causal/waiting links. Another requirement for the formalization of PMs is to capture the relationship between a CM and a PM when they are integrated in one DEMO model. In this article, we say that a Process Structure Diagram (PSD, a representation of a PM) is *matched* to an Organization Construction Diagram (OCD, a representation of CMs) as defined formally later in Definition 13.

We first propose an algebraic structure of a PSD and its submodels. In this step, we use the set-theoretic formalization and encode the structure of a DEMO PSD. It is because, as the earlier work [4] extensively discussed, other forms of formalism such as Z notation, temporal logic, description logics, and category theory turn out to be inappropriate for the purpose of the study. Those existing formalisms not only have different structures in modularity than DEMO PSDs, but are also inadequate for algebraic construction. In addition, it is beneficial to use the same type of formalism with the earlier work on CMs to leverage the knowledge and findings there. Although a PM specifies the state and transition space of the coordination world, it seems enough to rely on the set-theoretic formalism as long as we follow the metamodel. Secondly, an operation over the structure, the "merge" operation, is defined and associated properties are examined by formal proofs. Then, we consider a pair-wise merge operation for OCDs and PSDs. The result corroborates the coherence and consistency of DEMO models from a different angle. Towards the ultimate goal of constructing the algebra of DEMO models, including four aspect models, the current research project has worked on the algebra of PSDs. This article reports preliminary findings of the ongoing research, which captures a DEMO PSD as a static object.

The remainder of this paper is composed as follows. In the next chapter, we provide an overview of the background concepts and previous works related to the issue addressed in this article. The main result is given for the construction of an algebraic structure for DEMO PMs and an algebra over the structure, in Sects. 3 and 4, respectively. Section 5 discusses the contribution and limitations of this work, and concludes with a look at future directions.

2   Literature Review

2.1   Background Concepts

DEMO. Design & Engineering Methodology for Organizations (DEMO) is a modeling methodology for enterprises [2]. Although an enterprise—as an abstract term referring to (an assembly of) collaborative activities by human beings—has two perspectives (i.e., function and construction) in terms of systems engineering, DEMO spotlights the construction perspective. In addition to its focus on the construction of enterprises, DEMO also emphasizes the communications within enterprises, namely from the language/action perspective. For detailed information about DEMO, we refer the reader to [1,2].

DEMO is more suitable for the purpose and procedures of this research than other enterprise modeling languages, including business process modeling languages. It is because it provides a strong connection between the syntax and semantics. As shown in the rest of this article, elements of DEMO models are formalized by mathematics and studied from their isomorphic mathematical representations rather than the DEMO models themselves. Therefore, the strong connection is beneficial and essential to ensure the alignment between meanings (semantics) and symbols (syntax) when we encode DEMO models to mathematical representations and decode them back.

Aspect Models. The ontological model of an enterprise is represented by four aspect models, which each reflect a specific aspect of the enterprise. Because of the interest of this paper, we elaborate the Process Model here.

The Process Model (PM) of an enterprise specifies the state and transition space (i.e., the set of lawful sequences of the states) of the coordination world. Those states and transitions are largely specified by the universal transaction pattern. A PM is expressed in a Process Structure Diagram (PSD) for the whole and a Transaction Pattern Diagram (TPD) for each transaction kind. Although "exception(s)" dealt with in the model can be only specified in the TPD, the PSD occupies the central role in the PM. In other words, if we limit ourselves to a happy path, a PSD can be regarded as a PM without a TPD. Therefore, as a first attempt, this work concentrates on PSDs in the rest of this article. Figure 1a shows an example of a PSD. We use this metamodel as the grounds of the formalization of PSDs.

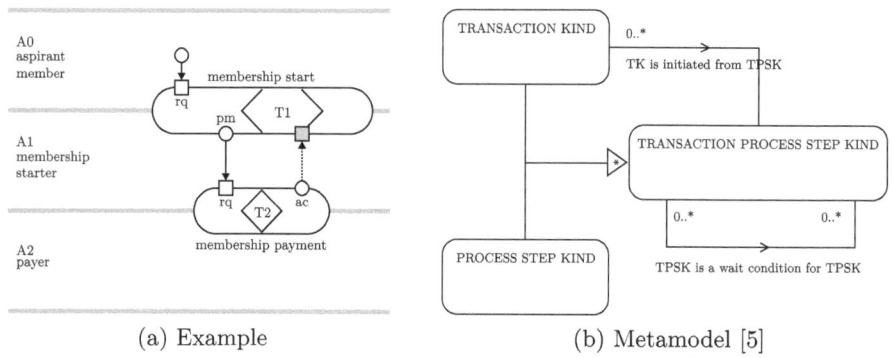

(a) Example   (b) Metamodel [5]

Fig. 1. DEMO PM

Abstract Algebra. Set theory and *structure* jointly play significant roles in mathematics. Intuitively, a *set* is simply a collection of objects known as elements that exist without any relation to each other. In contrast, a structure is a set together with some *relations* among elements of the set. Informally, such a structure is defined by specifying the elements and the relations. The rest of this section introduces an algebraic structure, which is one of well-known structures established by mathematicians.

Algebraic Structure. In abstract algebra, an algebraic structure consists of a set (called an underlying set), with one or more operations defined on the set, that satisfies certain axioms. The structure enables us to formalize and analyze objects that are more complex than just a collection of objects. Schematically, an algebraic structure is specified as a pair $\langle A; F \rangle$, where A is an underlying set and F is a family of finitary operations on A. Although we often assume the axioms of set theory, we may add or modify the axioms to shape the structure in order to represent more complex systems and allow derivations. A set is an instance of such an algebraic structure with F being empty (i.e., $\langle A; \emptyset \rangle$), whereas there are well-known variants such as groups, rings, lattices, and so on.

Algebra of Structures. Now we rise to a meta-level and introduce an algebra of (over) a structure. In the case of a set as a simple structure, *the algebra of sets* defines the laws over sets, the relations over sets (e.g., equality and inclusion), and set-theoretic operations over sets (e.g., union, intersection, and complement). Similarly, one may construct an algebra of an arbitrary structure. The ultimate goal of this research is to construct an algebra of DEMO models.

2.2 Related Works

Except for an earlier work [4], there exist very few formalizations for DEMO models, namely the CRISP model [6], Petri Nets [7,8], and XML schema [9]. Though we do not repeat the details of evaluations and insufficiency of those formalizations as a solution for the issue targeted by this research, the main deficiencies of the existing approaches are that none of them provide algebraic operations to form any type of algebra [4]. Since, this present work is dependent on the formalization of DEMO OCDs in the earlier work, we recall the essential definitions and theorem here. The full descriptions, including proofs and rationale, are available in [4].

Algebraic Structure of DEMO OCD. An algebraic structure of a DEMO OCD is defined by elements of sets and associated operations over the sets.

Definition 1 (Society). *A society $\mathbf{S} = \langle \mathbf{A}, \mathbf{T} \rangle$, as a universe, is made up of a set of actor roles $\mathbf{A}$ and a set of transaction kinds $\mathbf{T}$.*

Definition 2 (Actor Role). *An actor role has the following mappings:*

1. *$f_{ARno}: \mathbf{A} \to \mathbb{N}^0$ represents the number of each actor role, and*
2. *$f_{ARname}: \mathbf{A} \to \mathbf{string}$ represents the name of each actor role.*

Definition 3 (Transaction Kind). *A transaction kind t is a pair of actor roles (a, a'), where a and a' are called an initiator and an executor, respectively[1]. Formally, $t \in \mathbf{A} \times \mathbf{A}$. Transaction kinds have the following mappings:*

1. *$f_{Tno}: \mathbf{T} \to \mathbb{N}^+$ represents the number of each transaction kind,*
2. *$f_{Tname}: \mathbf{T} \to \mathbf{string}$ represents the name of each transaction kind,*
3. *$f_{Tin}: \mathbf{T} \to \mathbf{A}$ represents the initiator of each transaction kind,*
4. *$f_{Tex}: \mathbf{T} \to \mathbf{A}$ represents the executor of each transaction kind.*

Definition 4 (OCD). *Given that A is a subset of $\mathbf{A}$, and T is a subset of $\mathbf{T}$, an OCD is a pair $\langle A, T \rangle \in 2^{\mathbf{A}} \times 2^{\mathbf{T}}$ that satisfies the following conditions:*

Condition 1 (Unique Actor Role Name).
   $\forall a_i, \forall a_j \in A, (f_{ARname}(a_i) = f_{ARname}(a_j) \Rightarrow a_i = a_j)$
Condition 2 (Unique Transaction Kind Name).
   $\forall t_i, \forall t_j \in T, (f_{Tname}(t_i) = f_{Tname}(t_j) \Rightarrow t_i = t_j)$

---

[1] [4] assumes that there is only one initiator for each transaction kind.

Condition 3 (Numbering Convention).
$$\forall t \in T, f_{ARno}\left(f_{Tex}\left(t\right)\right) = f_{Tno}\left(t\right)$$
Condition 4 (Closure for Actor Role).
$$\forall t \in T, \forall a \in ActorRole, a \in f_{Tin}\left(t\right) \cup f_{Tex}\left(t\right) \Rightarrow a \in A$$
Condition 5 (Actor Role Participation).
$$\forall a \in A, \exists t \in T, a \in f_{Tin}\left(t\right) \cup f_{Tex}\left(t\right)$$

Based on this algebraic structure, we define the notion of "submodel" of a given OCD with the following formal specification.

Definition 5 (Sub-OCD). *Given an OCD $\langle A^\circ, T^\circ \rangle$, a couple $\langle A, T \rangle$ is said to be "a sub-OCD of $\langle A^\circ, T^\circ \rangle$" if $\langle A, T \rangle$ is an OCD (i.e., satisfies Definition 4), and $A \subseteq A^\circ$ and $T \subseteq T^\circ$ hold.*

Algebra of DEMO OCDs. Toward the algebra of DEMO OCDs, the previous work [4] introduced an operation over sub-OCDs of a given global OCD which performs the merger of the two sub-OCDs. Here we repeat the definition and its behavior as formulated in the form of theorems.

Definition 6 (OCD Merge). *Given an OCD $\langle A^\circ, T^\circ \rangle$ and its two sub-OCDs $\langle A_X, T_X \rangle$ and $\langle A_Y, T_Y \rangle$, the merge operation $\nabla : \left(2^A \times 2^T\right) \times \left(2^A \times 2^T\right) \to \left(2^A \times 2^T\right)$ is defined as $\langle A_X, T_X \rangle \nabla \langle A_Y, T_Y \rangle \triangleq \langle A_X \cup A_Y, T_X \cup T_Y \rangle$.*

Theorem 1. *The family of sub-OCDs of a given OCD is closed under the merge operation ∇.*

Theorem 2. *The merge operation ∇ is commutative and associative in the family of sub-OCDs of a given OCD.*

Relationship Between a CM and a PM. Although the specification of DEMO [5] specifies the metamodels for each aspect model separately, with less attention to the relationship between them, there are a few articles that address the relationship to ensure the integrity of the four aspect models. Particularly, we use [9] as a reference, which elaborates the requirements for transformation from a CM to a PM with coherence and consistency as "transformation rules".

3   Construction of Algebraic Structure

This section provides a formalization of the DEMO PSD and its submodels, which serves as the rigorous foundation for this research.

We use a common case study of Pizzeria [2] as a running example throughout this paper. For convenience, the PSD of Pizzeria is repeated in Fig. 2. Although it is a convention to draw actor roles with swim lanes, we omit them because they are not included in the metamodel of the PM.

3.1   PSD and Its Submodels

Algebraic Structure of PSD. As noted by the metamodel of the DEMO PM in Fig. 1b, the PSD has transaction kinds (TK), process step kinds (PSK), and transaction process step kinds (TPSK) as its components, with causal links and conditional links (denoted as "TK is an initiated from TPSK" and "TPSK is a wait condition for [another] TPSK", respectively). As a natural and straightforward approach, these components and links are formalized as sets and relations, respectively, below.

Let $\mathbf{T}$ be a set of transaction kinds. Transaction kinds have mappings to provide the name and number, namely $f_{Tname}\colon \mathbf{T} \to \mathbf{string}$ and $f_{Tno}\colon \mathbf{T} \to \mathbb{N}$. This definition is to give a universe, which contains all the entities we may wish to consider in a given situation. Thus, this set is not restricted to transaction kinds in a particular enterprise under consideration.

$\mathbb{I}$ is a set of process step kinds, which consists of Production-acts and Coordination-acts, namely $\mathbb{I} = \{\mathsf{rq}, \mathsf{pm}, \mathsf{ex}, \mathsf{st}, \mathsf{ac}\}$ in the case of the basic transaction pattern[2]. Since the static structure is of interest in the formalization, it is enough for $\mathbb{I}$ to be a set within the scope of this article. If the dynamic aspect in timing is considered, $\mathbb{I}$ should be accompanied by an order relation.

Based on the notations introduced so far, $(t, i) \in \mathbf{T} \times \mathbb{I}$ denotes a transaction process step kind, as specified in the metamodel by an *aggregation* of a transaction kind and a process step kind.

Example 1. Suppose we decide to consider Pizzeria as a given DEMO model. As it is a convention that sets and relations for the given global model are indicated by a circle ∘ in the superscript, the OCD of Pizzeria is denoted as $\langle \mathcal{A}^\circ, \mathcal{T}^\circ \rangle$, and the PSD is denoted as $\langle T^\circ, V^\circ, W^\circ \rangle$. In the scope of Pizzeria, we identify four transaction kinds: T1, T2, T3, and T4. We use t with the corresponding index to refer to these transactions kinds. Thus, the set of transaction kinds is $\mathbf{T} = \{\ldots, t_1, t_2, t_3, t_4, \ldots\}$. As clarified in Sect. 3.1, "..." is intentionally placed to express that $\mathbf{T}$ is a universe, and hence may include other transaction kinds.

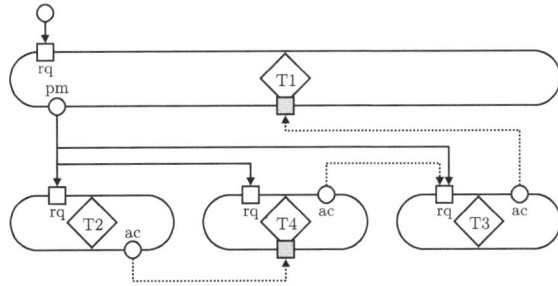

Fig. 2. PSD of Pizzeria

---

[2] When the formalization is expanded to the standard transaction pattern, $\mathbb{I}$ equals $\{\mathsf{rq}, \mathsf{pm}, \mathsf{ex}, \mathsf{st}, \mathsf{ac}, \mathsf{dc}, \mathsf{qt}, \mathsf{rj}, \mathsf{sp}\}$.

In the universe of transaction kinds, we create the set of transaction kinds for Pizzeria $T^\circ \subseteq \mathbf{T}$ as

$$T^\circ = \{t_1, t_2, t_3, t_4\}.$$

According to the names of the transaction kinds in Pizzeria, we have f_{Tname} : $t_1 \mapsto$ "completion", $t_2 \mapsto$ "preparation", $t_3 \mapsto$ "payment", $t_4 \mapsto$ "delivery". Based on this configuration, a transaction process step kind request of T1, for instance, is encoded into $(t_1, \mathsf{rq}) \in \mathbf{T} \times \mathbb{I}$.

Definition 7 (Activation Relation). *Let* $\mathbf{V}$ *be a relation over* $(\mathbf{T} \times \mathbb{I}) \times \mathbf{T}$, *where* $((t, i), t') \in \mathbf{V}$ *represents that a transaction kind* t' *is activated from a transaction process step kind* (t, i). *In this case,* $\mathbf{V}$ *is called an activation relation over* $\mathbf{T}$.

Example 2. The PSD of Pizzeria has three causal links from the promise of T1 to T2, T3 and T4. In our formalism these are expressed by the relation $\mathbf{V} \subseteq (\mathbf{T} \times \mathbb{I}) \times \mathbf{T}$. Thus, the relation for causal links in Pizzeria $V^\circ \subseteq \mathbf{V}$ is

$$V^\circ = \{((t_1, \mathsf{pm}), t_2), ((t_1, \mathsf{pm}), t_3), ((t_1, \mathsf{pm}), t_4)\}.$$

Definition 8 (Wait Relation). *Let* $\mathbf{W}$ *be a relation over* $(\mathbf{T} \times \mathbb{I}) \times (\mathbf{T} \times \mathbb{I})$, *where* $((t, i), (t', i')) \in \mathbf{W}$ *represents that a transaction process step kind* (t', i') *is a wait condition for a transaction process step kind* (t, i). *In this case,* $\mathbf{W}$ *is called a wait relation over* $\mathbf{T}$.

Example 3. The PSD of Pizzeria also specifies three conditional links from the accept of T2 to the execute of T4, from the accept of T4 to the request of T3, and from the accept of T3 to the execute of T1. These are encoded into the relation $\mathbf{W} \subseteq (\mathbf{T} \times \mathbb{I}) \times (\mathbf{T} \times \mathbb{I})$. In Pizzeria, $W^\circ \subseteq \mathbf{W}$ is

$$W^\circ = \{((t_2, \mathsf{ac}), (t_4, \mathsf{ex})), ((t_4, \mathsf{ac}), (t_3, \mathsf{rq})), ((t_3, \mathsf{ac}), (t_1, \mathsf{ex}))\}.$$

Just for the purpose of labeling, we define a *Transaction Process Step net (TPS-net)*, which is a triple-wise subset of a given tuple of sets as defined below.

Definition 9 (TPS-net). *A TPS-net is a triple* $\langle T, V, W \rangle$, *where* T *is a subset of* $\mathbf{T}$, V *is a subset of* $\mathbf{V}$, *and* W *is a subset of* $\mathbf{W}$.

Definition 10 (PSD). *Given that* T *is a subset of* $\mathbf{T}$, V *is a subset of* $\mathbf{V}$, *and* W *is a subset of* $\mathbf{W}$, *a PSD is a TPS-net* $\langle T, V, W \rangle$ *that satisfies*[3]

Property 1. $\forall t_i, t_j \in T, (f_{Tname}(t_i) = f_{Tname}(t_j) \implies t_i = t_j)$,
Property 2. $\forall ((t, i), t') \in V, t \in T \wedge t' \in T$, *and*
Property 3. $\forall ((t, i), (t', i')) \in W, t \in T \wedge t' \in T$.

---

[3] Although it is possible to impose more constraints such as "a wait condition must bridge two distinct transaction kinds", we aim for fidelity to the metamodel.

In plain English, the first property imposes the uniqueness of the names of transaction kinds as name equivalence; the second and third properties ensure the source and target transaction kinds of a link are included in the PSD. By definition, any PSD of $\langle T, V, W \rangle$ is also a TPS-net. As we assume the basic transaction pattern in this article, we do not explicitly write $\mathbb{I}$.

Example 4. It is easily confirmed that the tuple $\langle T^\circ, V^\circ, W^\circ \rangle$ satisfies the definition of a PSD.

Algebraic Definition of PSD Submodels. So far, we have described the formalism for PSDs. Here we proceed to define the notion of submodels for PSDs.

Definition 11 (Sub-TPS-net). *Given two TPS-nets $\langle T, V, W \rangle$ and $\langle T', V', W' \rangle$, $\langle T', V', W' \rangle$ is said to be a sub-TPS-net of $\langle T, V, W \rangle$ if $T' \subseteq T$, $V' \subseteq V$, and $W' \subseteq W$. Accordingly, we define "the family of sub-TPS-nets brought by a given PSD $\langle T, V, W \rangle$" that is the collection of all the TPS-nets $\langle T', V', W' \rangle$ that are sub-TPS-nets of $\langle T, V, W \rangle$. We use $\wp(\langle T, V, W \rangle)$ as a shorthand form for the family of sub-TPS-nets of $\langle T, V, W \rangle$.*

Definition 12 (Sub-PSD). *Given a PSD $\langle T, V, W \rangle$, a TPS-net $\langle T', V', W' \rangle$ is said to be a sub-PSD of $\langle T, V, W \rangle$ if $\langle T', V', W' \rangle$ is a sub-TPS-net of $\langle T, V, W \rangle$ and a PSD (i.e., it satisfies the three properties in Definition 10). Accordingly, we define "the family of sub-PSDs brought by a given PSD $\langle T, V, W \rangle$" that is the collection of all the PSDs $\langle T', V', W' \rangle$ that are sub-PSDs of $\langle T, V, W \rangle$. We use $\mathfrak{P}(\langle T, V, W \rangle)$ as a shorthand form for the family of sub-PSDs of $\langle T, V, W \rangle$.*

Example 5. Figure 3a illustrates one member of the family of sub-TPS-nets made by the given PSD. As the diagram notes, this is *broken* in the sense that the causal link from the promise of T1 to T2, but T2 is not included in Fig. 3a. Let $\langle T_0, V_0, W_0 \rangle$ denote this diagram. Since $T_0 = \{t_1, t_4\} \subseteq T^\circ$, $V_0 = \{((t_1, \mathsf{pm}), t_2), ((t_1, \mathsf{pm}), t_3)\} \subseteq V^\circ$, and $W_0 = \{((t_3, \mathsf{ac}), (t_1, \mathsf{ex}))\} \subseteq W^\circ$, $\langle T_0, V_0, W_0 \rangle$ is a sub-TPS-net of the given PSD according to Definition 11. However, this is not a sub-PSD of the given PSD due to the second property of Definition 12 not being satisfied. In contrast, Fig. 3b is a sub-PSD of the given PSD.

Although this definition requires a sub-PSD to satisfy the three conditions in Definition 10, we have to ensure only the second and third properties because of the following proposition. For readability, all proposition and theorem proofs are detailed in the Appendix.

Proposition 1. *Given a PSD $\langle T^\circ, V^\circ, W^\circ \rangle$, any member of the family of sub-TPS-nets $\wp(\langle T^\circ, V^\circ, W^\circ \rangle)$ satisfies the following condition:*

Property 1. $\forall t_i, t_j \in T, (f_{Tname}(t_i) = f_{Tname}(t_j) \implies t_i = t_j)$

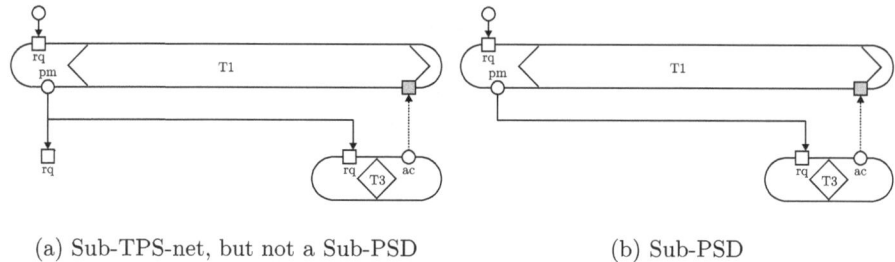

(a) Sub-TPS-net, but not a Sub-PSD   (b) Sub-PSD

Fig. 3. Example in Pizzeria

3.2   Match Between a CM and a PM

Independent of Sect. 3.1, this section describes another angle to restrict the freedom in a formalized PSD in terms of its coherence and consistency to the corresponding CM. Based on previous work [9, Chaps. 4 and 7], the requirements are shaped as (1) the set of transaction kinds in the PSD equals that in the OCD (up to name equivalence), and (2) the causal links between transaction kinds in the PSD should reflect the product structure of the OCD. These are formally defined as follows:

Definition 13 (Match). *Given an OCD* $\langle \mathcal{A}, \mathcal{T} \rangle$, *a PSD* $\langle T, V, W \rangle$ *is said to be* matched *to the OCD if*

1. $\mathcal{T} = T$, *and*
2. $\forall t, t' \in \mathcal{T}, (\exists a \in \mathcal{A}, f_{ex}(t) = a = f_{in}(t')) \iff (\exists i \in \mathbb{I}, ((t, i), t') \in V)$.

In this case, we may say an OCD and a PSD are matched.

The second condition above states the PSD is aligned to the product structure of the enterprise. Notably, based on our formalism, the product structure is uniquely derived from the OCD. Precisely speaking, the OCD can be uniquely transformed to the tree of organizational building blocks, and then uniquely converted to the tree of the product structure. According to the composition axiom [2], a root node is uniquely identified by a transaction kind that is initiated by the environmental actor role, or initiated by its executor itself (i.e., a self-activated transaction kind). In a simple case such as Pizzeria, the product structure is said to be a tree, which has one root transaction. The product structures of more complex cases, including Library [2], must be a collection of trees with sharing. The directed adjacency relation between transaction kinds, written as $\mathcal{R} \subseteq \mathcal{T} \times \mathcal{T}$, is uniquely obtained as $\mathcal{R} = \{(t, t') \mid \exists a \in \mathcal{A}, f_{ex}(t) = a = f_{in}(t')\}$.

4   Algebra of PSDs

This section introduces an operation and explores its properties. As the first step, we define the "merge" operation for two sub-PSDs of a given PSD. Then, in association with the merge operation of the OCD proposed in the earlier work (see Sect. 2.2), we elaborate the behavior of the DEMO OCD and PSD during the "merge" operation.

4.1   Merge Operation on Sub-PSDs

Analogous to set-theoretic union, the merge operation takes two sub-PSDs of a given PSD and produces an output that contains the two input models—specifically, the two sub-PSDs shall be a part (submodel) of the output, as formulated below.

Definition 14 (PSD Merge). *Given a PSD $\langle T^\circ, V^\circ, W^\circ \rangle$ and its two sub-PSDs $\langle T_X, V_X, W_X \rangle$ and $\langle T_Y, V_Y, W_Y \rangle$, the merge operation $\sqcup : \left(2^{\mathbf{T}} \times 2^{\mathbf{V}} \times 2^{\mathbf{W}} \right) \times \left(2^{\mathbf{T}} \times 2^{\mathbf{V}} \times 2^{\mathbf{W}} \right) \rightarrow \left(2^{\mathbf{T}} \times 2^{\mathbf{V}} \times 2^{\mathbf{W}} \right)$ is defined as*

$$\langle T_X, V_X, W_X \rangle \sqcup \langle T_Y, V_Y, W_Y \rangle \triangleq \langle T_X \cup T_Y, V_X \cup V_Y, W_X \cup W_Y \rangle .$$

Example 6. Finally, we demonstrate the merge operation using the sub-PSDs (Figs. 4a and b) of the given PSD in Fig. 2 to obtain the result in Fig. 4c. The sub-PSD in Fig. 4a is encoded as $\langle T_1, V_1, W_1 \rangle = \langle \{t_1, t_3\}, \{((t_1, \mathsf{pm}), t_3)\}, \{((t_3, \mathsf{ac}), (t_1, \mathsf{ex}))\} \rangle$ and in Fig. 4b as $\langle T_2, V_2, W_2 \rangle = \langle \{t_1, t_2\}, \{((t_1, \mathsf{pm}), t_2)\}, \emptyset \rangle$. Then, the merged PSD in Fig. 4c is obtained by Definition 14 as $\langle T_1, V_1, W_1 \rangle \sqcup \langle T_2, V_2, W_2 \rangle = \langle \{t_1, t_3\} \cup \{t_1, t_2\}, \{((t_1, \mathsf{pm}), t_3)\} \cup \{((t_1, \mathsf{pm}), t_2)\}, \{((t_3, \mathsf{ac}), (t_1, \mathsf{ex}))\} \cup \emptyset \rangle = \langle \{t_1, t_2, t_3\}, \{((t_1, \mathsf{pm}), t_3), ((t_1, \mathsf{pm}), t_2)\}, \{((t_3, \mathsf{ac}), (t_1, \mathsf{ex}))\} \rangle$.

Based on this definition, we claim three notable properties. First, the result of the operation is indeed a sub-PSD of the given PSD. Furthermore, this operation frees users from concerns about the order of the input models; changing the order of inputs does not change the result. These properties are formulated in theorems.

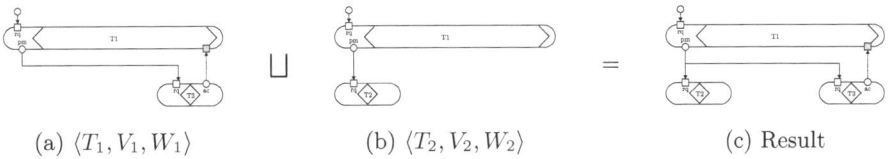

(a) $\langle T_1, V_1, W_1 \rangle$ (b) $\langle T_2, V_2, W_2 \rangle$ (c) Result

Fig. 4. PSD merge in Pizzeria

Theorem 3. *The family of sub-PSDs of a given PSD $\langle T^\circ, V^\circ, W^\circ \rangle$ is closed under the merge operation $\sqcup$, i.e., $\forall \langle T_X, V_X, W_X \rangle, \langle T_Y, V_Y, W_Y \rangle \in \mathfrak{P} (\langle T^\circ, V^\circ, W^\circ \rangle),$*

$$\langle T_X, V_X, W_X \rangle \sqcup \langle T_Y, V_Y, W_Y \rangle \in \mathfrak{P} (\langle T^\circ, V^\circ, W^\circ \rangle) .$$

Theorem 4. *The merge operation $\sqcup$ is commutative and associative on the family of sub-PSDs of a given PSD $\langle T^\circ, V^\circ, W^\circ \rangle$, i.e., $\forall \langle T_X, V_X, W_X \rangle$, $\langle T_Y, V_Y, W_Y \rangle, \langle T_Z, V_Z, W_Z \rangle \in \mathfrak{P}(\langle T^\circ, V^\circ, W^\circ \rangle)$,*

$$\langle T_X, V_X, W_X \rangle \sqcup \langle T_Y, V_Y, W_Y \rangle = \langle T_Y, V_Y, W_Y \rangle \sqcup \langle T_X, V_X, W_X \rangle$$

and

$$(\langle T_X, V_X, W_X \rangle \sqcup \langle T_Y, V_Y, W_Y \rangle) \sqcup \langle T_Z, V_Z, W_Z \rangle$$
$$= \langle T_X, V_X, W_X \rangle \sqcup (\langle T_Y, V_Y, W_Y \rangle \sqcup \langle T_Z, V_Z, W_Z \rangle).$$

4.2   Preserved Match Between the OCD Merge and PSD Merge

In addition to the behavior of the PSD merge as discussed so far, we claim the final theorem in this article guarantees the consistency between the PSD merge and OCD merge, as illustrated in Fig. 5.

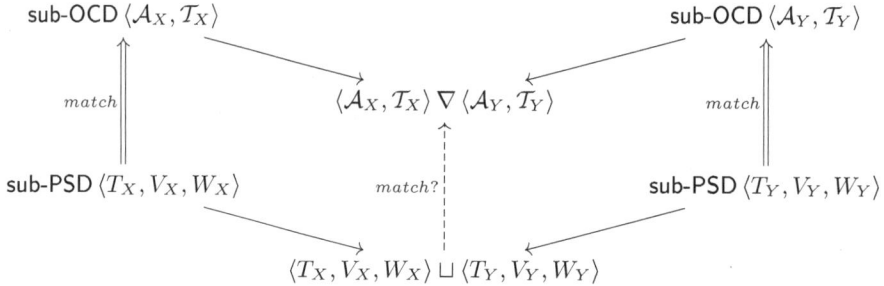

Fig. 5. Illustration of Theorem 5

Theorem 5. *Given an OCD $\langle \mathcal{A}^\circ, \mathcal{T}^\circ \rangle$ and a matched PSD $\langle T^\circ, V^\circ, W^\circ \rangle$, let $\langle \mathcal{A}_X, \mathcal{T}_X \rangle$ and $\langle \mathcal{A}_Y, \mathcal{T}_Y \rangle$ be sub-OCDs of the given OCD and $\langle T_X, V_X, W_X \rangle$ and $\langle T_Y, V_Y, W_Y \rangle$ be sub-PSDs of the given PSD.*

If $\langle T_X, V_X, W_X \rangle$ is matched to $\langle \mathcal{A}_X, \mathcal{T}_X \rangle$ and $\langle T_Y, V_Y, W_Y \rangle$ is matched to $\langle \mathcal{A}_Y, \mathcal{T}_Y \rangle$, the result of the PSD merge is matched to the result of the OCD merge, i.e., $\langle T_X, V_X, W_X \rangle \sqcup \langle T_Y, V_Y, W_Y \rangle$ is matched to $\langle \mathcal{A}_X, \mathcal{T}_X \rangle \nabla \langle \mathcal{A}_Y, \mathcal{T}_Y \rangle$.

Practitioners may take advantage of this theorem. A project to merge two sub-OCDs and sub-PSDs can be divided into four smaller tasks as follows: (1) ensure one sub-PSD is matched to one sub-OCD, (2) ensure the other sub-PSD is matched to the other sub-OCD, (3) merge the two sub-OCDs, and (4) merge the two sub-PSDs. The four tasks can be completed separately by different individuals and/or computers without collaboration. Moreover, nobody has to check whether the result of task 4 (PSD merge) is matched to the result of task 3 (OCD merge) because it is guaranteed by this theorem.

5   Conclusion and Future Research

The goal of this paper was to present a straightforward construction of an algebraic structure for DEMO PSDs and explore the merge operation as an instance of model manipulations. The primary result states that the family of sub-PSDs of a given PSD is closed under the merge operation, while preserving the integrity of sub-OCDs and sub-PSDs. Furthermore, the merge operation is commutative and associative in that family. Stated another way, the finding may reinforce the coherence and consistency of DEMO from a different angle.

The contribution of this work is not limited to the theoretical aspect, is but of significance to practitioners as well. The commutativity and associativity of the merge operation ensure the same result regardless of the order of the operations. If one wishes to merge three models, A, B, and C but B is not available yet, one can first merge A and C, then merge the result and B later when B becomes available. Moreover, Sect. 4.2 exhibits the contribution of this work in terms of division of labor in a practical scenario. As we mentioned in Sect. 1, these formal specifications of DEMO aspect models and investigation of mathematical behavior should provide a solid foundation of computer-aided design environments, which may play an important role in the future development of Way of Supporting, which has attracted less attention so far, but will be of significance in Enterprise Engineering.

Although this research has answered the issue addressed in Sect. 1, there is an intrinsic limitation in that the formalization only captures the static aspect of DEMO PMs. Considering each transaction kind is a finite state machine (FSM) in the sense of the universal transaction pattern, a PM is a composition of FSMs. Thus, this paper ignores dynamic aspects such as the behavior of the composite FSMs and may miss some important points. In fact, the authors noticed that a PSD in Fig. 6b is intuitively and realistically a good submodel of a given PSD in Fig. 2 in the sense that the accept of T4 is a waiting condition for the execute of T1, particularly when you interpret the purpose of making submodels is to ignore one or more specific transaction kinds. However, the PSD in Fig. 6b is invalid in the presented formalization because the waiting condition in question is not included in the given global PSD. This observation implies that the formalization could be improved, probably by revising the definition of *part of* to reflect the dynamic aspects. This limitation also implies more case studies are required for further validation.

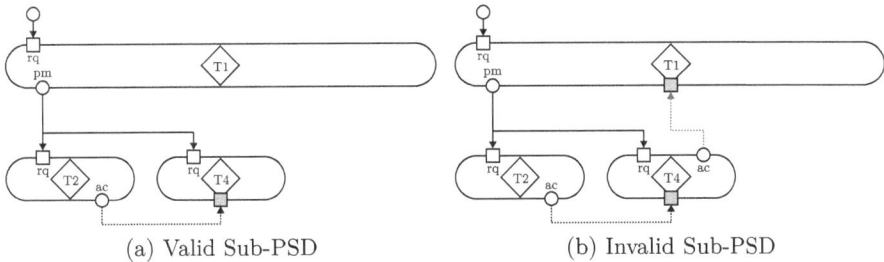

(a) Valid Sub-PSD   (b) Invalid Sub-PSD

Fig. 6. Two Sub-PSDs

Future research plans include three stages. The first is to improve the formalization of PSDs to reflect the dynamic aspects, for instance, by introducing behavioral preorders and equivalence such as traces, failures, and (bi)similarity of finite state machines instead of set-theoretic ones. Furthermore, the proposed artifact could be validated through further case studies involving real organizations. Although these two steps have higher priority, the third one is to ultimately expand the formalization by covering other operations such as set-theoretic intersection and complement to complete the algebra of DEMO PSDs. Nevertheless, the authors are convinced this contribution is an abstract but important step in evolving the research and development of Way of Supporting in engineering enterprises.

Acknowledgments. This work was supported in part by Program for Leading Graduate Schools "Academy for Co-creative Education of Environment and Energy Science", MEXT, Japan.

Appendix: Proofs

Proof of Proposition 1. Suppose $\langle T, V, W \rangle$ is an arbitrary member of the family of sub-TPS-nets $\wp \left(\langle T^\circ, V^\circ, W^\circ \rangle \right)$. Considering that $\langle T^\circ, V^\circ, W^\circ \rangle$ is a PSD, for any t_i and t_j in T°, if $f_{Tname}(t_i) = f_{Tname}(t_j)$ then $t_i = t_j$. Now that $T \subseteq T^\circ$ by Definition 9, for any t_i and t_j in $T \subseteq T^\circ$, if $f_{Tname}(t_i) = f_{Tname}(t_j)$ then $t_i = t_j$.

Proof of Theorem 3. Since $\langle T_X, V_X, W_X \rangle$ and $\langle T_Y, V_Y, W_Y \rangle$ are members of the family of sub-PSDs of a given PSD $\langle T^\circ, V^\circ, W^\circ \rangle$, we have $T_X \subseteq T^\circ$, $T_Y \subseteq T^\circ$, $V_X \subseteq V^\circ$, $V_Y \subseteq V^\circ$, $W_X \subseteq W^\circ$, and $W_Y \subseteq W^\circ$ by Definition 9. Then, because $T_X \cup T_Y \subseteq T^\circ$, $V_X \cup V_Y \subseteq V^\circ$, and $W_X \cup W_Y \in W^\circ$ hold, it is obvious with Definition 14 that $\langle T_X, V_X, W_X \rangle \sqcup \langle T_Y, V_Y, W_Y \rangle$ is a sub-TPS-net of $\langle T^\circ, V^\circ, W^\circ \rangle$. Next, for any $((t, i), t')$ in $V_X \cup V_Y$, if $((t, i), t')$ is in V_X [resp. V_Y], $t \in T_X$ and $t' \in T_X$ [resp. $t \in T_Y$ and $t' \in T_Y$] holds, hence $t \in T_X \cup T_Y$ and $t' \in T_X \cup T_Y$. Thus, any $((t, i), t')$ in $V_X \cup V_Y$ satisfies the second property of Definition 10. Similarly, any $((t, i), (t', i'))$ in $W_X \cup W_Y$ satisfies the second property of Definition 10. Note that the first property of Definition 10 is satisfied by Proposition 1 because $\langle T_X, V_X, W_X \rangle \sqcup \langle T_Y, V_Y, W_Y \rangle$ is a TPS-net. Hence, $\langle T_X, V_X, W_X \rangle \sqcup \langle T_Y, V_Y, W_Y \rangle$ is a PSD. Therefore, by Definition 12, $\langle T_X, V_X, W_X \rangle \sqcup \langle T_Y, V_Y, W_Y \rangle$ is a sub-TPS-net of $\langle T^\circ, V^\circ, W^\circ \rangle$ and a PSD, thus a sub-PSD of the given PSD $\langle T^\circ, V^\circ, W^\circ \rangle$.

Proof of Theorem 4. The commutativity and associativity of merge operation $\sqcup$ are obvious from those of set-theoretic union of sets.

Proof of Theorem 5. Since $\langle T_X, V_X, W_X \rangle$ is matched to $\langle \mathcal{A}_X, \mathcal{T}_X \rangle$ and $\langle T_Y, V_Y, W_Y \rangle$ is matched to $\langle \mathcal{A}_Y, \mathcal{T}_Y \rangle$, Definition 13 gives $\mathcal{T}_X = T_X$, $\mathcal{T}_Y = T_Y$, "$\forall t, t' \in \mathcal{T}_X, (\exists a \in \mathcal{A}_X, f_{ex}(t) = a = f_{in}(t')) \iff (\exists i \in \mathbb{I}, ((t, i), t') \in V_X)$", and "$\forall t, t' \in \mathcal{T}_Y, (\exists a \in \mathcal{A}_Y, f_{ex}(t) = a = f_{in}(t')) \iff (\exists i \in \mathbb{I}, ((t, i), t') \in V_Y)$".

Thus, $\mathcal{T}_X \cup \mathcal{T}_Y = T_X \cup T_Y$. It also holds that $\forall t, t' \in \mathcal{T}_X \cup \mathcal{T}_Y, (\exists a \in \mathcal{A}_X \cup \mathcal{A}_Y,$ $f_{ex}(t) = a = f_{in}(t')) \iff (\exists i \in \mathbb{I}, ((t, i), t') \in V_X \cup V_Y)$. Therefore, by Definition 13, the PSD of $\langle T_X, V_X, W_X \rangle \sqcup \langle T_Y, V_Y, W_Y \rangle$ is matched to the OCD $\langle \mathcal{A}_X, \mathcal{T}_X \rangle \nabla \langle \mathcal{A}_Y, \mathcal{T}_Y \rangle$.

References

1. Dietz, J.L., Hoogervorst, J.A., Albani, A., Aveiro, D., Babkin, E., Barjis, J., Caetano, A., Huysmans, P., Iijima, J., van Kervel, S.J., Mulder, H., Op't Land, M., Proper, H.A., Sanz, J., Terlouw, L., Tribolet, J., Verelst, J., Winter, R.: The discipline of enterprise engineering. Int. J. Organisational Design Eng. **3**(1), 86–114 (2013)
2. Dietz, J.L.G.: Enterprise Ontology: Theory and Methodology. Springer, Heidelberg (2006)
3. Seligmann, P., Wijers, G., Sol, H.: Analyzing the structure of IS methodologies - an alternative approach. In: Maes, R. (ed.) Proceedings of the First Dutch Conference on Information Systems, Amersfoort, pp. 1–28 (1989)
4. Suga, T., Iijima, J.: Does 'Merging DEMO Models' satisfy the associative law? - Validation of partial models and merge operation. In: Proceedings of the 7th International Joint Conference on Knowledge Discovery, Knowledge Engineering and Knowledge Management, Lisbon, vol. 2, pp. 467–478. SCITEPRESS - Science and Technology Publications (2015)
5. Dietz, J.L.G.: DEMO specification language (version 3.3, November 2015) (2015). http://www.ee-institute.org/download.php?id=165&type=doc. Accessed 15 Jan 2017
6. Dietz, J.L.G.: The DELTA theory - understanding systems. Technical report TR-FIT-15-05, Czech Technical University in Prague (2015)
7. Barjis, J.: Automatic business process analysis and simulation based on DEMO. Enterp. Inf. Syst. **1**(4), 365–381 (2007)
8. Fatyani, T., Iijima, J., Park, J.: Transformation of DEMO model into coloured petri net: Ontology based simulation. In: 6th International Conference on Knowledge Engineering and Ontology Development, KEOD 2014, Italy, pp. 388–396. SCITEPRESS (Science and Technology Publications, Lda.) (2014)
9. Wang, Y.: Transformation of DEMO models into exchangeable format. Master's thesis, Delft University of Technology (2009)

A DEMO Machine - A Formal Foundation for Execution of DEMO Models

Marek Skotnica[1(✉)], Steven J.H. van Kervel[2], and Robert Pergl[1]

[1] Czech Technical University, Prague, Czech Republic
skotnicam@gmail.com, robert.pergl@fit.cvut.cz
[2] Formetis, Boxtel, The Netherlands
steven.van.kervel@formetis.nl

Abstract. The discipline of enterprise engineering and the DEMO methodology provide enterprise designers with a formal techniques to design companies where competency, responsibility and authority is clearly defined. In such companies, process-based anomalies can be avoided and people tend to cooperate more effectively and contentedly.

These techniques are so far mostly used just for business process modeling consultancy. DEMO-based software systems are needed to adopt and support these techniques in professional companies. This paper proposes a theoretical computation concept called DEMO Machine that provides us with formal foundations for a simulation of DEMO models. We demonstrate these formal foundations on a Volley Club example.

Keywords: DEMO machine · Enterprise engineering · DEMO simulation · DEMO software implementation

1 Introduction

The Enterprise engineering community has been working on formal theories and methodologies for more than 15 years. The results were found to surpass the state of the art of business process management (BPM) approaches in terms of formal correctness, ontological completeness, and anomalies [1]. But, so far an adoption of these principles in practice is very slow. One of the reasons is that the largest benefit from these theories is provided to middle-sized or large companies and these organizations tend to change very slowly. In addition, a new technology adoption is associated with high risks. Large IT systems with many complex features are required, as well, usually provided by large companies such as IBM, Pega, Oracle, or Microsoft. There are no such large DEMO-based IT systems so far. As argued in the FAR Ontology paper [2], it is not easy to understand how the DEMO models are simulated. This work builds on van Kervel's work [3], simplifies it according to the Occam's law and enables for further extensions. It also builds on ForMetis company professional experience in building DEMO-based systems.

The goal of this paper is to propose a theoretical computation foundations that are easy to understand (like BPMN) and yet allow to express all the DEMO

© Springer International Publishing AG 2017
D. Aveiro et al. (Eds.): EEWC 2017, LNBIP 284, pp. 18–32, 2017.
DOI: 10.1007/978-3-319-57955-9_2

aspect models. This is a prerequisite for building DEMO-based IT systems that could compete with state-of-the art BPM systems (BPMS). For this purpose, we propose a DEMO Machine – an abstract formalism, which can be used for DEMO model simulation and DEMO model code implementation.

The paper is organized as follows: In Sect. 2, the research question is summarized. In Sect. 3, the underlying scientific foundations are briefly discussed. In Sect. 4, formal definitions of DEMO Machine are proposed, investigated, and represented in a formal notation. In Sect. 5, the proposed theories are demonstrated on a Volley Club example. In Sect. 6, the current results are summarized and further research is proposed.

2   Research Question

This paper elaborates on a research question proposed in FAR Ontology paper Sects. 3.1 and 3.2 [2]. The DEMO Machine is meant as a formal computation model (similar to the e.g. the Turing Machine). The DEMO Machine needs to take into account challenges that are induced by the execution level and thus not addressed in DEMOSL [2]. The research question was stated as: **"How should a DEMO Machine be designed to interpret DEMOSL?"**.

3   Theories Used and Related Work

Theories used in this paper were already mentioned in the FAR Ontology paper [2], therefore we just offer a brief summary of them: Guizzardi's ontology theories [4], the Enterprise Ontology [5], the DEMO methodology [5], van Kervel's papers [3,6].

This paper is also influenced by related work in this area, most notably: Figueira and Aveiro [7], Huysmans [8], Krouwel [9], and Op't Land [10].

4   DEMO Machine

This section elaborates on the research question proposed in Sect. 2. To investigate characteristics of a software system, it is better to do it on its formal model rather than on its software implementation. We do take an inspiration from Turing's invention called the Turing Machine [11], which was the first universal computer made in 1936, years before any physical computers existed.

DEMOSL provides specification for the DEMO models of an enterprise. However, for the simulation of the models there are no definitions provided yet. Therefore, we define the missing concepts and propose a formal DEMO Machine that is able to simulate the models. This machine is independent on any software implementation, and it is only based on the mathematical concepts.

4.1   DEMO Model Definitions

In this section, essential DEMO model definitions are provided in a form that is suitable for DEMO Machine simulation. The semantics of these concepts is aligned with the DEMO theory [5]. DEMO model is an ontological representation of an enterprise. Demo models are commonly represented by four aspect diagrams – OCD, PSD, OFD, and AM. Diagrams together express a DEMO model. The following formalization deals with the DEMO model itself.

Definition 1. *Actor Role. An actor role is an ordered tuple:*

$$ActorRole := (Identifier, ActorRoleType) \tag{1}$$

Identifier – A unique identifier of an actor role.
ActorRoleType $\in \{Elementary, Composite\}$

An elementary actor role is an atomic amount of authority, responsibility, and competence. It is a producer in exactly one transaction, and a customer of zero, one, or more transactions [5]. A composite actor role is a network of transaction kinds and (elementary) actor roles, of which one does not (want to) know the details [12].

Definition 2. *Transaction Kind. A transaction kind is an ordered tuple:*

$$Transaction := (Identifier, TransactionKindName, Executor, Initiators) \tag{2}$$

The second axiom of the Ψ-theory states that coordination acts are performed as steps in universal patterns [5]. These patterns, also called transactions, always involve two actor roles and are aimed at achieving a particular result [5]. These patterns are formally defined in Sect. 4.3.

Definition 3. *Causal Link. A causal link is an ordered tuple:*

$$\begin{aligned} CausalLink := (&SourceTransactionKind, SourceState, \\ &TargetTransactionKind, TargetState, MinCardinality, \\ &MaxCardinality, InitiatorActorRole) \end{aligned} \tag{3}$$

InitiatorActorRole – An initiator Actor Role to distinguish to which executor this link applies since a transaction can have multiple initiators.

According to the theory, a causal link is defined as: *"a link between a coordination act and its resulting coordination fact, indicating the fact is result of the act."* [12]. A Causal link is used in a tree-like structure to define a business process composed of multiple transactions. For example, when there is a causal link from T1/pm to T2/rq it means that you can initiate a new T02 instance from a T01 instance that is in state promised or a later state.

Definition 4. *Conditional Link.* *A conditional link is an ordered tuple:*

$$ConditionalLink := (SourceTransaction, SourceState, TargetTransaction,$$
$$TargetState, InitiatorActorRole)$$

$$(4)$$

InitiatorActorRole – An initiator Actor Role to distinguish to which executor this link applies since a transaction can have multiple initiators.

Conditional link restricts the source transaction state from being reached until the causal link's cardinalities are satisfied. For example, there is a causal link from T1/pm to T2/rq with cardinality 1..1. There is a conditional link from T02/ac to T01/st. This means that you can perform cAct T01/st only when one child transaction T02 reached ac.

Definition 5. *DEMO Model.* *A DEMO Model is an ordered tuple:*

$$DEMOModel := (Identifier, TransactionKinds, ActorKinds,$$
$$ConditionalLinks, CausalLinks, Facts, Rules)$$

$$(5)$$

A DEMO Model is a conceptual representation of an enterprise or a sub-enterprise at a given time frame. Facts and rules definitions are provided in [2].

4.2   DEMO Enterprise Application Definitions

We considered model definitions so far, but once the simulation of a model takes place, the instances need to be taken into the account because they represent the day to day operation of an enterprise.

Definition 6. *Enterprise Position.* *A DEMO enterprise position is an ordered tuple:*

$$EnterprisePosition := (Identifier, ActorRoles) \qquad (6)$$

Identifier – Is and identifier of DEMO Enterprise Position
ActorRoles – Is a finite set of actor roles. An ActorRole can belong to several Enterprise Positions.

An enterprise position is defined as a coherent set of actor roles. In practice, it means a principle to group these roles and define responsibilities, competence, and authorities at a generic level; e.g. sales director, production manager etc. These are sometimes also called *functional roles*.

Definition 7. *Actor.* *An actor is an ordered tuple:*

$$Actor := (Identifier, EnterprisePositions) \qquad (7)$$

Identifier – A unique identifier of an actor instance.
EnterprisePositions – A finite set of enterprise positions

An actor is a person or group of persons (board) that operates in an enterprise in given enterprise positions.

Definition 8. *Transaction*. *A transaction is an ordered tuple:*

$$Transaction := (DEMOModel, TransactionKind, ParentTransaction, \\ InitiatorActor, ExecutorActor, State) \quad (8)$$

DEMOModel – A model according which a transaction behaves.
TransactionKind – Is a type of transaction.
ParentTransaction – Is parent transaction. May be empty in case of a root transaction.
InitiatorActor – An actor that initiated the transaction.
ExecutorActor – An actor that is responsible for the execution side of the transaction. May be empty or changed over time as the execution responsibility may be delegated.
State – The current transaction state. States are further explained in Sect. 4.3.

A transaction represents an actual situation, in which the transaction kind is carried out (by people).

Definition 9. *DEMO Enterprise Application*. *A DEMO enterprise application is an ordered tuple:*

$$DEMOEnterpriseApplication := (Identifier, PublishedModels, \\ EnterprisePositions, Actors, Transactions) \quad (9)$$

PublishedModels – Is a finite set of DEMO models.
Identifier – Is an identifier of DEMO Enterprise Application.
EnterprisePositions – A finite set of enterprise positions that actors can participate in. Enterprise positions can only contain actor roles defined in PublishedModels.
Actors – Is a finite set of Actors.
Transactions – Is a finite set of Transactions.

A DEMO enterprise application represents an actual enterprise that consists of DEMO models, actors, and their interactions. A DEMO model is a conceptualization of an enterprise in one given time. A real-world enterprise changes over time, and therefore it needs to act according to multiple DEMO models, resp. their versions. For example, a mortgage company creates a contract in 1990 based on certain conditions, and these conditions do still need to apply in 2017 even though conditions for new mortgages are different. A DEMO Enterprise can also consist of multiple sub-enterprises or departments represented by multiple DEMO models. This concept is very important for a software system implementation, it allows aggregation of agenda – work-items from multiple DEMO Model instances.

4.3   DEMO Axiom Definitions

There are three DEMO axioms that need to be formalized and performed in order to calculate an agenda for a given transaction instance. Agenda, and cAct definitions are provided in [2].

Definition 10. *DEMO Axiom is a function that takes an agenda and calculates a set of cActs:*

$$DEMOAxiom : (Transaction, Agenda) \rightarrow \{cAct\} \qquad (10)$$

Transaction Axiom. For the purposes of software simulation, we do formally define the transaction axiom as a state machine in Fig. 1. The circles are states (cFacts), and the boxes are allowed actions (cActs). With this state machine, an implementation of Transaction DEMOAxiom function is straightforward.

We propose also some practical changes that make the transaction axiom suitable for building enterprise information systems. We added a possibility to start a transaction instance without being requested. This supports the real-world situations where people start to negotiate about a transaction. Documents are created, but no request has been made, yet. A distinction whether a transaction starts with a request or initiate or both is done as an extra information on the causal link in the PSD.

The second deviation from the theory is that we do not support revoke of all states at all times. This simplification is mostly because of the composition axiom. For example, when a child transaction is created from the promised state, therefore you are not able to revoke the promise. Revokes combined with the composition axiom are quite a challenging topic in the execution and are a subject to further and mostly empirical research.

Composition Axiom. Composition axiom adds cActs based on the conditional and causal links, so that the transaction instances can form a process. An implementation of this axiom is out of scope of this paper, and it is a subject for further research.

Rule Axiom. Rule axiom adds cActs based on the conditional and causal rules based on the definitions from the FAR Ontology [2].

4.4   DEMO Model Simulation

An Operation of an organisation is the manifestation of its construction in time [13]. Simulation is the imitation of the operation of a real-world process or system over time [14]. DEMO model simulation is the imitation of the operation of a DEMO model for a purpose of validation of the model correctness. DEMO model execution is a DEMO model simulation for the purpose of supporting an operation of an enterprise IT system.

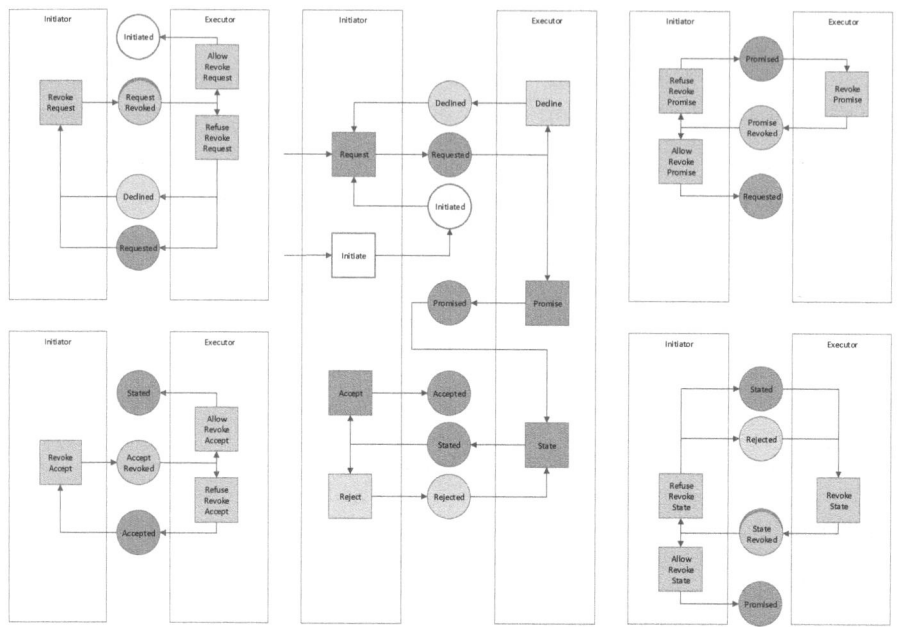

Fig. 1. Transaction axiom state machine

Definition 11. *A **DEMO Machine** is an ordered tuple:*

$$DEMOMachine := (DEMOEnterpriseApplication,$$
$$ExternalFactImplementations, TransactionInstanceLinking, \qquad (11)$$
$$InputInstructions, OutputMessages)$$

DEMOEnterpriseApplication – A DEMO enterprise application.
TransactionInstanceLinking – Ternary relation that represents connections between transaction instances in the outside world.
ExternalFactImplementations – Outside world implementations of functions that calculate external facts.
InputInstructions – A set of instructions that the machine needs to process.
OutputMessages – Results produced by the machine that represent facts about a behaviour of an enterprise.

The DEMO Machine is receiving instructions on the input and producing messages on the output.

The list of allowed instructions is:

- **GetActorAgenda(Actor)** – Writes an *Agenda* for a specified *Actor* into *OutputMessages.*
- **PerformCAct(cAct)** – Performs a *cAct* and puts a new Agenda for the actor instance (defined in cAct) into *OutputMessages.* Performing an empty cAct causes a recalculation of the model instance.

The Algorithm 1 shows a pseudo-code of how the agenda is calculated for a transaction instance.

Algorithm 1. Agenda calculation

1: **function** CALCULATEAGENDA(transactionInstance, actorPerformCActs)
2:     #Adds actors perform cActs
3:     $agenda \leftarrow$ actorPerformCActs
4:     #Adds allowed cActs based on Transaction axiom
5:     $agenda.add(TransactionAxiom(agenda))$
6:     #Adds allowed and restricted cActs based on Composition axiom
7:     $agenda.add(CompositionAxiom(agenda))$
8:     #Adds perform and restricted cActs based on Rule axiom
9:     $agenda.add(RuleAxiom(agenda))$
10:     #Find perform cActs that are allowed and not restricted.
11:     **if** agenda has cAct c to perform **then**
12:         #Performs cActs selected to be performed
13:         $PerformCAct(transactionInstance, c)$
14:         #Transaction states have been changed so recalculation of agenda is needed.
15:         **return** $CalculateAgenda(transactionInstance, nil)$
16:     **else**
17:         #No cActs to be performed found, agenda reached a stable state.
18:         **return** $agenda$

The presented algorithm is just a high-level abstract schema. A detailed description of the DEMO Machine calculation is outside of the scope of this paper.

5   Proof of Concept – Volley Club

In this section, a proof-of-concept DEMO Machine is demonstrated on a Volley club model from the book "The Essence of the Organization" by Jan Dietz [15]. The model is well specified in the book, so we do not elaborate on it much, and we rather point out the differences in our approach and the proposed way of simulation.

To verify the formal definitions, we created a proof-of-concept software implementation of the presented DEMO Machine. In this section we use a general object-oriented pseudo-code inspired by C# to implement the simulation according to the definitions provided above.

5.1   DEMO Model

The organization construction diagram (OCD) in Fig. 2 contains two transactions describing the situation where a customer comes into the club, requests a membership, pays for it, and he becomes a member.

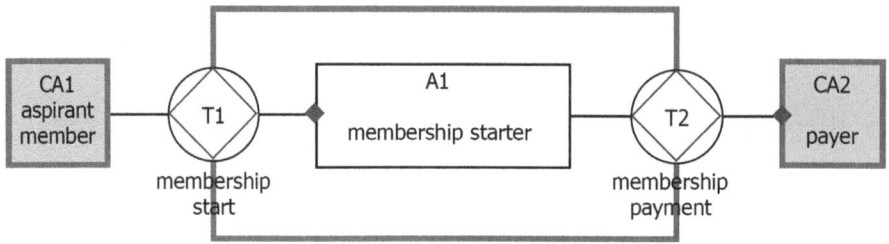

Fig. 2. OCD model volley club [15]

The Process Diagram (PSD) describes how are the two transactions related. The membership payment is requested after a membership start is promised. There is also a conditional link which specifies that the membership execution phase can't be done until the membership payment is accepted. Cardinality is not mentioned here, but we expect only one payment per membership. Later payments are not part of the model.

The Action Model (AM) here consists of four rules, and all of them are for the membership starter (A1). The logic of working with facts defined in the OFD is also included in the rules, but DEMO materials do not elaborate on how they should be dealt with. The precise definition, how to execute this AM rules, is also not provided, but for a communication between human stakeholders, this notation is sufficient.

1. **Action Rule for A1(1)** – When the membership start (T1) is requested, in case the person who is requesting is eligible, then it is automatically promised, otherwise declined. Eligibility means that the person is old enough, starting day of the membership is the first day of some month and maximum number of members was not reached.
2. **Action Rule for A1(2)** – When the membership start (T1) is promised, then automatically request the membership payment.
3. **Action Rule for A1(3)** – When the membership payment (T2) is stated, while the paid amount for the membership has been paid, then accept the membership payment (T2), otherwise reject (T2).
4. **Action Rule for A1(4)** – When the membership start (T1) is promised while the membership payment (T2) is accepted, then execute the membership start (T1) and state the membership start (T2).

5.2 DEMO Machine Model

Here is how the same Volley club model looks like when described by the concepts introduced in this paper.

OCD and PSD remain the same. They are represented as:

```
AspirantMember = ("Aspirant member", Composite);
MembershipStarter = ("Membership starter", Elementary);
Payer = ("Payer", Composite);
T1 = ("T01","Membership Start", MembershipStarter, {AspirantMember})
T2 = ("T02","Membership Payment", Payer, {MembershipStarter})
VolleyClubModel = ("Volley Club", {T1, T2}, {AspirantMember,
    MembershipStarter, Payer}, ...)
```

Information about memberships or person is likely to be stored in an external database and there is no use in duplicating them inside the DEMO Machine, as explained in [2].

The action model implementation differs from the DEMO, so let's go through the Volley club business rules and see how they are expressed in the DEMO Machine.

Action Rule for A1(1) is represented by an external fact and a causal rule. The external fact contains all the business conditions that are needed in order to evaluate, whether a person is eligible for a membership. The *LogicalProposition* is there merely to suggest what logic should be used to evaluate such fact. The real logic then lies in the outside world implementation, and it calls the database. A benefit of this approach is that we do not need to change the model when this business rule is modified. A new implementation version is simply plugged in, and the system goes on.

The causal rule *T1RequestedCausalRule* is there to implement the action part (state transition) of the AM rule. It says: "When an instance of *transaction1* is in state Requested and fact *IsMemberElegibleFact* is evaluated as True, then add a cAct with SettlementType=Perform and Intention=Promise to the transaction instance agenda. If the fact is evaluated as False, then add a cAct with SettlementType=Perform and Intention=Decline to the transaction instance agenda." This explanation may seem to be more complicated than the previous action rule, but it covers much more scenarios. Adding of an enforcing cAct is used instead of a direct state transition, because the transition may be forbidden by some conditional rule. The state transition also needs to be allowed by the transaction or the composition axiom. In case of multiple rules enforcing different state transitions, a priority should be assigned to the rules.

```
IsMemberElegibleFact = ExternalFact("Is member eligible for application
    ?",
  LogicalProposition = "Person.Age >= Minimal_Required_Age",
  VolleyClubCalculationEngineId)
T1RequestedCausalRule = CausalRule(T1, Requested, IsMemberElegibleFact,
  cAct(T1, T1.Current, T1.Current.Executor, Promise, Perform),
  cAct(T1, T1.Current, T1.Current.Executor, Decline, Perform))
```

Action Rule for A1(2) – is represented by a causal rule and an external fact. The external fact will be always True in this case since there are no business conditions. The causal rule is expressed bellow and it says: "If the transaction instance of type T1 is in state Promised and the *TrueExternalFact* is evaluated as True, then add a cAct that (i) performs creation of a new instance of T2 that will be a child of the current T1 transaction instance and (ii) will be in state Created to the current transaction instance agenda".

```
T1_Promised_CausalRule = (T1, Promised, TrueExternalFact, cAct(T1, T1.
    Current, T1.Current.Executor, Create(T2, 1), Perform), null)
```

An interesting problem is that the transaction instance T1 can get into state Promised multiple times. Does it mean that it should create a new instance of T2 each time it gets there? And does it depend on some external system? In this model, the creation of unwanted transactions is controlled by the 1..1 cardinality defined in the PSD. However, for generic purposes, we introduced a possibility for external fact implementation to return a number of transactions to be created together with the fact result. This is the way, how we can control how many transactions are created.

Another problem is in determining the executor actor instance for a created transaction instance of T2. It is clear in this particular model that the membership payer will be the same person as an aspirant member. However, it is not formally defined. We do delegate this problem to the outside world implementation. Once it is notified about the created instance of T2, it has all the information it needs to assign the executor. More empirical experience shows, whether this is sufficient, or a more sophisticated solution needs to be designed.

This proof of concept implementation does not contain the composition axiom, and therefore the rules to create child transactions are not implemented, as well.

Action Rule for A1(3) – is represented by a causal rule and an external fact. The external fact is a business rule that determines whether the paid amount was enough. The causal rule then performs accept or reject.

```
IsPaidAmountEnoughFact = ExternalFact("Is paid amount for membership
    enough?",
  LogicalProposition = "this.Membership.AmountToPay <= this.Membership.
    Payment.AmountPaid", VolleyClubCalculationEngineId)
T2StatedCausalRule = CausalRule(T2, Requested, IsPaidAmountEnoughFact,
  cAct(T1, T2.Current, T2.Current.Executor, Accept, Perform),
  cAct(T2, T2.Current, T2.Current.Executor, Reject, Perform))
```

Action Rule for A1(4) – is represented by a conditional rule and a communication fact. We only want the execution phase to be allowed when the child transaction of T1 instance is in state Allowed. We capture such fact using a communication fact that says: "Are all current transaction instance children with type T2 accepted?". If there is no child transaction with type T2, the fact is evaluated as Undefined.

The conditional rule says: "If there is a cAct with Intention=State and SettlementType=Allow within the current transaction instance agenda and the fact *IsMembershipPaidFact* is not evaluated as true, then a cAct with Intention=State and SettlementType=Restrict is added to the current transaction agenda." Simply put, the transaction instance state Stated can be only reached when the fact is True.

```
IsMembershipPaidFact = CommunicationFact("Is membership paid?",
   CommunicationFactExpression = "this.children<T02>.all(t => t.state ==
      accepted)", VolleyClubCalculationEngineId)
T2StatedCausalRule = ConditionalRule(T1, IsMembershipPaidFact, State)
```

5.3   Volley Club Outside World Implementation

The outside world consists of the implementation of external facts, transaction relation provider, and state change receiver. It can be implemented in any programming language, as long as it provides values required in the definitions. In our proof of concept implementation, we created a simple implementation of such system that accessed a database and returned relevant values. However, a detailed description of such implementation is not relevant for purposes of this paper.

5.4   Step by Step Execution

In this section, we will provide detailed description of what happens in the execution of Volley club model during the happy-flow scenario.

At first, a Volley club enterprise application is created, and an implementation of the outside world is attached. The Activity Log shows all the changes in the running enterprise application, and we present all steps of the simulation bellow.

Step 1 – We create enterprise positions and attach them to actor roles. Then we assign actors to enterprise positions. Marek is going to be a Customer, which is an enterprise position with actor roles *Aspirant member* and *Payer*. Elisabeth is going to be an Employee – the Membership starter since she works in the Volley club.

Step 2 – Marek would like to be a member of Volley club, so he initiates a new transaction 1 instance and selects its executor to be Elisabeth. He is prepared to do a request of the membership, but he needs to fill out the starting day. He fills today and performs the request. Because there is nothing to restrict Marek's request, the transaction moves to state Requested. New *Membership* object is created, and it stores the data Marek entered. In state T1 Requested, a causal rule is defined and therefore evaluated. Marek is 27 years old, and that is enough to be a member of Volley club. The causal rule adds enforcing cAct to the agenda, and it moves the transaction to state Promised. In the Promised state, there is a conditional rule that restricts the State from being performed before the membership is paid. The communication fact is evaluated as Undefined, because there is no accepted child T2. No interaction was required from Elisabeth.

```
New transaction T01 was created with name=T01.1.
T01.1:Request:Allow
Initiator of T01.1 performed Request.
```

```
T01.1:Request:Allow,T01.1:Request:Perform
Fact "Is member eligible for application?" was evaluated as True.
T01.1:Promise:Allow,T01.1:Decline:Allow,T01.1:RevokeRequest:Allow,T01.1:
    Promise:Perform
Fact "Is membership paid?" was evaluated as Undefined.
T01.1:State:Allow,T01.1:RevokePromise:Allow,T01.1:State:Restrict
```

Step 3 – Elisabeth received a request from Marek, and she would like to deliver him the membership. However, she needs to ask for a payment first, and therefore she initiates a new transaction 2. After the transaction 2 was initiated, the conditional rule was evaluated again. Now, the result of communication is not Undefined but False. This is because the T2 exists.

```
New transaction T02 was created with name=T02.2.
Fact "Is membership paid?" was evaluated as False.
T01.1:State:Allow,T01.1:RevokePromise:Allow,T02.2:Request:Allow,T01.1:
    State:Restrict
```

Step 4 – Elisabeth calculated a membership fee for Marek, and she requested a membership payment. The communication fact is still False.

```
Initiator of T02.2 performed Request.
Fact "Is membership paid?" was evaluated as False.
T01.1:State:Allow,T01.1:RevokePromise:Allow,T02.2:Request:Allow,T01.1:
    State:Restrict,T02.2:Request:Perform
Fact "Is membership paid?" was evaluated as False.
T01.1:State:Allow,T01.1:RevokePromise:Allow,T02.2:Promise:Allow,T02.2:
    Decline:Allow,T02.2:RevokeRequest:Allow,T01.1:State:Restrict
```

Step 5 – Marek promises to pay for the membership. Before he states the payment, he needs to fill the amount to pay based on the requested amount created by Elisabeth. He fills 30 Euro and states the payment. When transaction 2 is stated, a causal rule that validates if the paid amount is valid is activated. The sum of money matches and transaction 2 is accepted. Communication fact "Is membership paid?" is finally evaluated as True.

```
Executor of T02.2 performed Promise.
Fact"Is membership paid?" was evaluated as False.
T01.1:State:Allow,T01.1:RevokePromise:Allow,T02.2:Promise:Allow,T02.2:
    Decline:Allow,T02.2:RevokeRequest:Allow,T01.1:State:Restrict,T02.2:
    Promise:Perform
Fact "Is membership paid?" was evaluated as False.
T01.1:State:Allow,T01.1:RevokePromise:Allow,T02.2:State:Allow,T02.2:
    RevokePromise:Allow,T01.1:State:Restrict
Executor of T02.2 performed State.
Fact "Is membership paid?" was evaluated as False.
T01.1:State:Allow,T01.1:RevokePromise:Allow,T02.2:State:Allow,T02.2:
    RevokePromise:Allow,T01.1:State:Restrict,T02.2:State:Perform
Fact "Is paid amount for membership enough?" was evaluated as True.
Fact "Is membership paid?" was evaluated as False.
```

```
T01.1:State:Allow,T01.1:RevokePromise:Allow,T02.2:Accept:Allow,T02.2:
    Reject:Allow,T02.2:RevokeState:Allow,T02.2:Accept:Perform,T01.1:State
    :Restrict
Fact "Is membership paid?" was evaluated as True.
T01.1:State:Allow,T01.1:RevokePromise:Allow,T02.2:RevokeAccept:Allow
```

Step 6 – Elisabeth is allowed to state the membership, and she does so. The communication fact "Is membership paid?" was evaluated once more because transaction 2 could have changed in the meantime.

```
Executor of T01.1 performed State.
Fact"Is membership paid?" was evaluated as True.
T01.1:State:Allow,T01.1:RevokePromise:Allow,T02.2:RevokeAccept:Allow,T01
    .1:State:Perform
T01.1:Accept:Allow,T01.1:Reject:Allow,T01.1:RevokeState:Allow,T02.2:
    RevokeAccept:Allow
```

Step 7 – Marek accepts the membership creation.

```
Initiator of T01.1 performed Accept.
T01.1:Accept:Allow,T01.1:Reject:Allow,T01.1:RevokeState:Allow,T02.2:
    RevokeAccept:Allow,T01.1:Accept:Perform
T01.1:RevokeAccept:Allow,T02.2:RevokeAccept:Allow
```

Step 8 - Marek is a proud member of Volley club. We can see that his record was created in the database. The *TransactionId* is there to associate the DEMO engine transaction instance identifier with the membership record. The relation could be also stored inside the DEMO engine as transaction instance's external identifier.

6   Conclusions and Further Research

In this paper, we proposed a theoretical computation model called the DEMO Machine, and we demonstrated its capability to simulate DEMO models on a Volley club example. We strive to contribute to developing model-driven systems based on DEMO models. However, there are still many topics for further research. Apart from the specific topics mentioned in the text, we would like to stress the evolvability of DEMO models and its consequences, alignment with existing business process management systems, and adoption of DEMO-based systems for the end users, so they are easy to use and comprehend.

Acknowledgement. This research has been supported by CTU SGS grant No. SGS16/120/OHK3/1T/18.

References

1. Nuffel, D., Mulder, H., Kervel, S.: Enhancing the formal foundations of BPMN by enterprise ontology. In: Albani, A., Barjis, J., Dietz, J.L.G. (eds.) CIAO!/EOMAS -2009. LNBIP, vol. 34, pp. 115–129. Springer, Heidelberg (2009). doi:10.1007/ 978-3-642-01915-9_9

2. Skotnica, M., Kervel, S.J.H., Pergl, R.: Towards the ontological foundations for the software executable DEMO action and fact models. In: Aveiro, D., Pergl, R., Gouveia, D. (eds.) EEWC 2016. LNBIP, vol. 252, pp. 151–165. Springer, Cham (2016). doi:10.1007/978-3-319-39567-8_10
3. Van Kervel, S.J.H.: Ontology driven enterprise information systems engineering. TU Delft, Delft University of Technology (2012)
4. Guizzardi, G.: Ontological foundations for structural conceptual models, vol. 015. University of Twente, Enschede (2005)
5. Dietz, J.L.G.: Enterprise Ontology Theory and Methodology. Springer, Heidelberg (2006)
6. Van Kervel, S., Dietz, J., Hintzen, J., Van Meeuwen, T., Zijlstra, B.: Enterprise ontology driven software engineering. In: Proceedings of the 7th International Conference on Software Paradigm Trends, ICSOFT 2012, pp. 205–210 (2012).
7. Figueira, C., Aveiro, D.: A new action rule syntax for DEmo MOdels based automatic worKflow procEss geneRation (DEMOBAKER). In: Aveiro, D., Tribolet, J., Gouveia, D. (eds.) EEWC 2014. LNBIP, vol. 174, pp. 46–60. Springer, Cham (2014). doi:10.1007/978-3-319-06505-2_4
8. Huysmans, P., Oorts, G., Bruyn, P., Mannaert, H., Verelst, J.: Positioning the normalized systems theory in a design theory framework. In: Shishkov, B. (ed.) BMSD 2012. LNBIP, vol. 142, pp. 43–63. Springer, Heidelberg (2013). doi:10.1007/978-3-642-37478-4_3
9. Krouwel, M.R., Op 't Land, M.: Combining DEMO and normalized systems for developing agile enterprise information systems. In: Albani, A., Dietz, J.L.G., Verelst, J. (eds.) EEWC 2011. LNBIP, vol. 79, pp. 31–45. Springer, Heidelberg (2011). doi:10.1007/978-3-642-21058-7_3
10. Op't Land, M.: Exploring normalized systems potential for dutch MoD's Agility (2011). Accessed 25 April 2014
11. Turing, A.M.: On computable numbers, with an application to the entscheidungsproblem. Proc. London Math. Soc. **s2–42**(1), 230–265 (1937)
12. Dietz, J.L.: The Essence of Organization - an Introduction to Enterprise Engineering. Sapio bv (2012)
13. Jan, D., Jan, H.: Theories in Enterprise Engineering Memorandum - TAO
14. Banks, J., Carson, J.S., Nelson, B.L., Nicol, D.M.: Discrete-Event System Simulation, 3rd edn. Prentice Hall, Upper Saddle River (2000)
15. Dietz, J.L.G.: Enterprise ontology - understanding the essence of organizational operation. In: Chen, C.S., Filipe, J., Seruca, I., Cordeiro, J. (eds.) Enterprise Information Systems VII, pp. 19–30. Springer, Dordrecht (2006)

Standards and Laws

Adding Quality of Information
to the Ontological Model of an Enterprise

Ron Deen[1,4](✉), Johan Mijs[2,4], and Martin Op 'T Land[3,4]

[1] Ilionx, Hondiuslaan 40, 3528 AB Utrecht, The Netherlands
rdeen@ilionx.com
[2] Cultuurconnect, Priemstraat 51, 1000 Brussel, Belgium
johan.mijs@cultuurconnect.be
[3] Capgemini Netherlands, P.O. Box 2575, 3500 GN Utrecht, The Netherlands
Martin.OptLand@capgemini.com
[4] Antwerp Management School, Sint-Jacobsmarkt 9-13, 2000 Antwerp, Belgium

Abstract. Critical to the success of an enterprise is to not only remember and share information, but also to make sure it meets the required quality. We developed a method for adding information quality requirements to the ontological DEMO model of an enterprise, by first defining information products, second determine relevant quality characteristics from the *ISO/IEC 25012:2008 Data Quality Model for Software product Quality Requirements and Evaluation* standard, and third add these quality characteristics to the ontological model [1]. As a benefit we found that it not only offers a systematic way to determine the needed quality of information to support the business organization, it also reveals a way to model this need on an ontological level through the creation, remembering and recalling of new information and by the use of this new information in the action rules of new or existing actor roles. Further research is required to ascertain this is a good starting point for eliciting software requirements to support responsibilities regarding quality of information.

Keywords: Enterprise ontology · Quality of information · ISO/IEC 25012:2008 · Data quality

1 Introduction

Enterprises are increasingly dependent on information systems to support the delivery of services and products to their customers [17]. In order for information systems to offer effective support in the process of information delivery, their software requirements should be aligned with the information need of the business organization. In literature a lot of different approaches can be found about how to design and engineer information systems, and how to elicit requirements for such systems. We are proponents of approaches that first model the business requirements (requirements regarding the services and products that

© Springer International Publishing AG 2017
D. Aveiro et al. (Eds.): EEWC 2017, LNBIP 284, pp. 35–49, 2017.
DOI: 10.1007/978-3-319-57955-9_3

are delivered to customers), before modeling the information need of the business organization and designing the information systems that fulfill that need. The Design and Engineering Methodology for Organizations or DEMO [6] is a methodology that enables the modeling of the business organization of an enterprise and its information need in a coherent way. We believe this is the right starting point for requirements analysis and system modeling for a supporting information system.

We have studied several other approaches, and agree with Van Lamsweerde [12] in his preface that some books on the subject mainly consider the requirements engineering *process* and discuss general principles, guidelines and documentation formats, while other books focus on *software* design. Whitten and Bentley [20] explain traditional modeling approaches like structured analysis and information engineering, and also the Unified Modeling Language (UML), which is a more modern object oriented approach. Sommerville [18] discusses the requirements engineering process, as well as system modeling and exclusively uses UML as a modeling notation. All these approaches are focused on software design, and are completely without concepts to model the business organization that will be using the software system. The approach of Van Lamsweerde is better in our view, in the sense that he starts from a Goal orientation, which enables the modeling of the goals of the business organization. His method eventually drills down from goals to requirements for a supporting software system. This makes it possible to trace back which goals have led to which system functionality.

Although DEMO allows for the modeling of the information need of the business organization, currently it lacks a proper treatment of the quality of the needed information. It refers to non functional requirements in the Generic System Development Process (GSDP), equalling non functional requirements with the constructional requirements that are guiding the constructional design of a system [6]. However, how such requirements are added to a constructional design remains unclear. When considering quality of information requirements specifically, ISO/IEC 25012:2008 lists 15 data quality characteristics which can be applied to data and information [7].

Quality of information is never an end in itself. It must always support the business requirements of an enterprise. We propose a method that starts from the essential model of the enterprise and has the following steps: 1. Define the information products from the essential model of the (or part of the) enterprise. 2. Determine the relevant information quality characteristics based on the business requirements of the enterprise. 3. Model quality of information in the ontology of the B and I-organization of the enterprise. Further research is required to ascertain that this more complete or enriched ontological model allows for the elicitation of software requirements that better take information quality into account.

The remainder of this paper is structured as follows. In Sect. 2 we describe our research approach and problem statement. In Sect. 3 we describe the method we created in more detail. In Sect. 4 we show how, by using this method, quality of information requirements can be added to the essential model of an enterprise. We finally demonstrate and evaluate the practical relevance of our method in the real life case of a Unified Library System for Flemish public libraries in Sect. 5. We end with the conclusions of our research in Sect. 6 and do some suggestions for further research.

2   Research Approach

2.1   Problem Statement

Enterprises are goal oriented cooperatives to be implemented by people and means [14]. The business organization of an enterprise is responsible for the business function, which means it is responsible for delivering the services and products to customers. The requirements that need to be taken into account while designing the business function we will call business requirements. While executing its business function, the business organization depends on information delivered by the information organization. Poor quality of information leads to customer dissatisfaction, increased operational cost, less effective decision making, a reduced ability to make and execute strategy and, more subtle, poor information quality hurts employee morale, breeds organizational mistrust, and makes it more difficult to align the enterprise [17]. In order for information systems to offer effective support in this process of information delivery, their software requirements should be aligned with the information need of the business.

The Design and Engineering Methodology for Organizations or DEMO [4] claims to offer a coherent, comprehensive, consistent, and concise method for modeling organizations. The distinctive property of organizations is that the active elements are human beings in their role of social individual. DEMO models every organization in three distinct aspect organizations: a business, information, and documental organization. The business organization produces the goods and services to the environment. The information organization supports the business organization with information services and the documental organization supports the information organization with documental services. Applying DEMO leads to a so called ontological model which is free from implementation choices. With implementation we understand that technological means like information technology, machines, or even humans according to Dietz in [4], are assigned to transactions and actor roles. Dietz defines the *essential model* of an enterprise as the ontological model of the business organization, including the specification of its information need [5].

Currently DEMO theory lacks a proper treatment of quality characteristics, although we found a reference to nonfunctional requirements in the Generic System Development Process (GSDP) which is described in [6]. The GSDP is a general framework for understanding the process of design, where an object system is designed (and further developed) to be used by another (the using) system. Two types of design are part of the GSDP. The first is functional design, the second constructional design, which is supported by constructional requirements, which we will call non functional requirements in this text (Fig. 1).

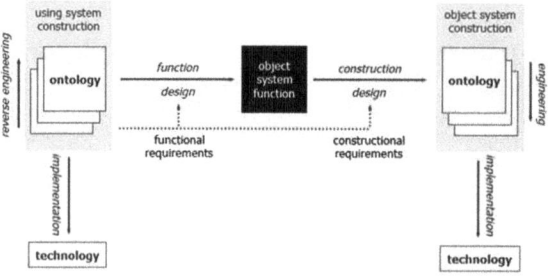

Fig. 1. Generic system development process

Although nonfunctional requirements are mentioned in GSDP, it remains unclear how they fit into the constructional design of a system. Since the highest level constructional model is the ontological model, it seems logical to first investigate the impact of nonfunctional requirements on the ontological model of the object system.

In literature many taxonomies of nonfunctional requirements can be found. An overview of classifications of nonfunctional requirements for system analysis and design is given by Adams who introduces a taxonomy of 27 nonfunctional requirements in 4 categories: design, viability, sustainability and adaptation [2]. Van Lamsweerde [12] classifies nonfunctional requirements into quality of service (QoS), compliance, architecture constraints and development constraints. Op 'T Land and Proper further subclassify quality of service for the three DEMO aspect organizations into quality of business (QoB), quality of information (QoI) and quality of data (QoD) and position them in the Enterprise Engineering Framework (EEF), which is inspired by xAF and GSDP [15]. From the previous discussion we conclude that business requirements can be split into functional business requirements and nonfunctional or quality of business (QoB) requirements. Functional business requirements are about what services and products are offered to customers. Nonfunctional or quality of business requirements are about how services and products are offered to the customers, they specify the need for a certain quality to make the business function useful.

In Fig. 2 we have visualized the scope of our research project. The yellow shaded area is our starting point. First, it contains the function perspective of the B-organization, which consists of the business content (the products or services themselves, e.g. the home-delivery of a pizza) and the quality of business (e.g. home-delivery within half an hour). Although literature often positions quality of service as non-functional, in EEF it is positioned in the function perspective, because of its black-box observability. Second, it contains the essential model (the ontological construction perspective of the B-organization and the information content of the Function perspective of the I-organization). Note that choices made for implementing the B-organization can also lead to information requirements for the design of the I-organization (single yellow square), but this was out of scope of our research project.

			System type		
	Context		General		
P			Business	Informational	Datalogical
e	**Function**	Content	Business content	Information content	Data content
r					
s		Quality of Service	Quality of Business (QoB)	Quality of Information (QoI)	Quality of Data (QoD)
p					
e	**Construction**	Ontological			
c					
t		Implementation	Parties & people			
i					
v			Means			
e					

Fig. 2. Positioning our research project (pink section) in EEF

The pink shaded area is the subject of our research. First, it includes the function perspective of the I-organization, because we want to find functional and information quality requirements. Second, it includes the ontological construction perspective of the B-organization as well as that of the I-organization, because we want to add the functional and quality requirements that are found to the design of the ontologies of the B and I-organization. The Document-organization, or D-organization, is mainly out of scope, but we did include an example of a D-actor role and a D-transaction in the section *Artefact demonstration and evaluation*.

To summarize the above in a *Problem statement*: When you want to use DEMO to create an essential model of (part of) the enterprise as the starting point for eliciting requirements for an information system, it is not clear how to take quality of information requirements into account.

This leads us to the following *Research question*: How do we systematically add information quality requirements to the essential model of an enterprise, given that its functional and nonfunctional business requirements are known?

2.2 Proposed Method: Design Science

The research method that seemed most appropriate for our purposes was design science. Recker states that the fundamental principle of design science research is that knowledge and understanding of a design problem and its solution are acquired in the building and application of an artefact [16]. This is exactly what we wanted to do. By actually designing a method for adding information quality requirements to the ontological model of an enterprise and by trying out this method, we tried to gain more insight in how information quality fits in the methodological concepts of DEMO. By carrying out our research in the described manner we adhered to the three core criteria for design science mentioned by Recker:

1. *The artefact is novel.*
   To our knowledge there is no prior research about how quality of information can be added to the essential DEMO model of an enterprise.
2. *The artefact is useful (makes a positive difference compared to existing work).*
   Alternative methods of requirements engineering that we studied, either focus on the engineering process, or focus on software design, without introducing concepts to model the business organization that will use the information system. Our method explicitly uses requirements of the business organization to find the relevant nonfunctional requirements and adds these requirements to the essential model. We believe this makes its easier to make the transition to lower level construction models that ultimately could lead to the implementation of an information system.
3. *The usefulness of the artefact is proved.*
   We have tested the practical usefulness of the artefact in a real-life case.

Recker provides a short introduction to design science using the work of Hevner ([10]), connecting environment, research and a knowledge base through a relevance, design and rigor cycle. Further exploration of the literature on design science led us to Johannesson and Perjons who developed a method framework to create an overview of a design science project: the design science canvas [9]. The canvas helped us to get an structured and concise visual overview of our research project.

3   Artefact Requirements

Our research artefact is a method that is intended to be used by enterprises who are looking for a structured way to improve the quality of the information that is used to support their business organization. To be able to use the method effectively, the following conditions have to be met for the part of the enterprise that is within scope:

1. The essential DEMO model has been created.
2. The functional and nonfunctional business requirements are defined.

To let enterprises benefit optimally from our method, first of all it should consider a complete list of information quality characteristics. We took the quality characteristics from ISO/IEC 25012:2008 as a starting point [7]. This standard lists 15 quality characteristics which can be applied to data. The standard makes a difference between *inherent data quality* as the degree to which quality characteristics of data have the intrinsic potential to satisfy stated and implied needs when data is used under specified conditions. *System dependent data quality* is defined as the degree to which data quality is reached and preserved within a computer system when data is used under specified conditions. In our method, we have only used the inherent quality characteristics, leaving out the system (and implementation) dependent characteristics Availability, Portability and Recoverability.

Second, to let our method be hands-on and scientifically sound we positioned it within the Total Data Quality Management (TDQM) method of Wang [19]. Wang applies Total Quality Management from the field of product manufacturing to information products. He defines a cycle to continuously Define, Measure, Analyze and Improve information quality, based on Demings Plan, Do, Check, Act cycle. Our approach fits in the Define phase of TDQM, but doesnt touch the other steps.

Third, the method is based on supporting the business requirements of the enterprise, which means in short that the need for information to have a certain quality characteristic depends on the requirements the business has.

Fourth, the underlying aim of the method is to make it possible to design implementation construction models that take business requirements and the required information quality into account.

4   Artefact

In order to demonstrate our method we have made use of the well known DEMO reference case Pizzeria [4]. DEMO consists of four integrated models for modeling the business organization. The most concise model is the Organization Construction Model (Fig. 3).

This model shows the actor roles of the business organization, the production acts they perform and the transactions they initiate to put other actor roles to work. It also shows on a high level what is the functional information need of

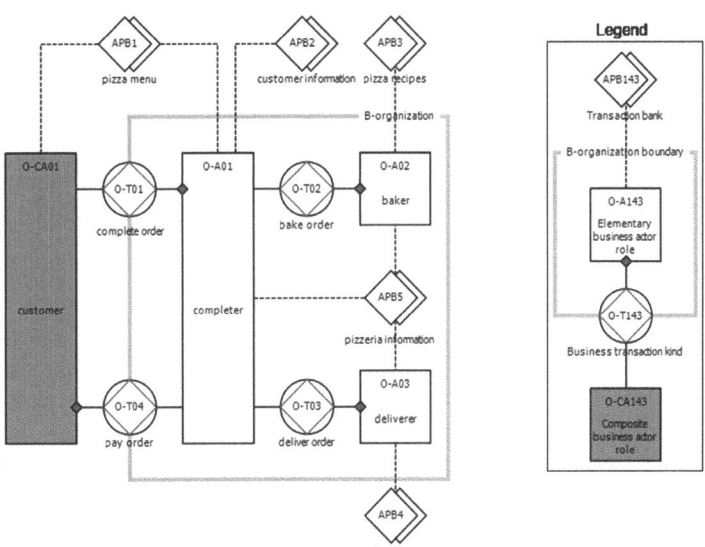

Fig. 3. Organization construction diagram of the B-organization of pizzeria

each actor role, by means of information links to so-called transaction banks containing information.

Next to the Construction Model, there are three other models. The Process Model shows how actors coordinate the process of delivering a good or service to the environment. The Fact Model shows the information that is relevant for the operation of the enterprise. The most comprehensive model is the Action Model. It specifies all of the above, but adds the rules that actor roles use to decide how they will act on tasks they are requested to perform (their agenda). For instance, when an actor role receives a request to produce some goods, he will use an action rule to decide if he will promise or decline the request. In the following paragraphs the steps of our method are performed, to end up with a more elaborate ontological model, taking quality of information into account.

4.1 Define Information Products from the Essential Model of the (or Part of the) Enterprise

The first step of our method consists of defining the information products of the Pizzeria. Wang defines information products as *information conceptualized in terms of its functionalities for information consumers.* In DEMO the information consumer is the B-actor role. The specific need of the B-actor role on a specific moment is best expressed in the Action Model, in which each Action Rule specification expresses how the B-actor role should react on a specific agendum. How the B-actor will react not only depends on the rule itself, but also on the information that is used to calculate the outcome of the rule. Therefore we use the following definition of information product: *a piece of information that is part of an Action Rule specification and that is used by the B-actor role to decide about his reaction on an agendum.* We don't make any demands about the granularity of the information products in Action Rule specifications.

So, to find all information products, it is sufficient to create the Action Model of the ontological model of our scope of interest. How to do this is elaborately described in [4]. To visualize the information products and the way they can be retrieved, we have used Construction Diagrams to show all relevant I-transactions. As a convention information products are I-transactions that are initiated by a B-actor role.

Although we have defined the complete set of information products for the Pizzeria case in the master thesis [1], we will show one example here. The specific part we will look at is the support of the B-actor role *baker* during execution of transaction *O-T02 bake order*. Figure 4 shows the relevant information products in a Construction Diagram, namely *Share order* and *Derive order baking instructions*. One can imagine that to be able to bake the ordered pizzas, information about the order and information about how to bake the pizzas is required.

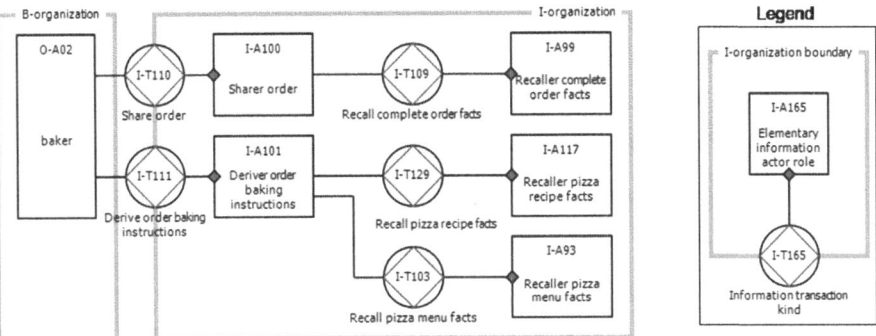

Fig. 4. I-organization supporting the baker in execution phase of O-T02 bake order

4.2   Determine the Relevant Quality of Information Characteristics Based on the Business Requirements of the Enterprise

To be able to apply quality characteristics sensibly we first need to gather the business requirements that apply to our scope of interest. Second, we need to prioritize the information products we have found in step 1, based on their relevance for the business requirements. Third, we need to determine which quality characteristics of the relevant information products should be improved, by questioning information producers (the I-organization) and the information consumers (the B-organization) about the importance of each quality characteristic, about the perceived and expected level of quality in a characteristic.

In our example we will use the following business requirement: *Our pizzas have a unique taste and texture.* A big part of the customer base specifically orders the pizzas of Pizzeria because of the unique taste and texture. For management it is very important that these unique properties cannot be copied by other pizzerias. For baking the pizzas, the *baker* uses the information product *Derive order baking instructions.* The management of Pizzeria wants to enforce that unauthorized personnel cannot access this information. Based on this business requirement we decide we need quality characteristic *Confidentiality* of the Data Quality Model standard ISO/IEC 25012. We adapted the definitions of the standard for a better match with DEMO terminology and because we wanted to explicitly use the term information product in the definition [1]. The ISO/IEC 25012 definition of Confidentiality is: *The degree to which data has attributes that ensure that it is only accessible and interpretable by authorized users in a specific context of use.* We changed the definition into: *The degree to which the facts of an information product are shared only with actors with the authority to perform a specific actor role.*

4.3   Model Information Quality Characteristics in the Ontology of the B and I-Organization of the Enterprise

In the last step of the method we will let the management of Pizzeria make decisions about how to add the selected quality of information requirements to the essential model of the Pizzeria. We don't know of any clearcut way to do this. The solution depends on the specific situation and the combined expertise of the modeling team.

The first moment that an employee receives authorization to perform some responsibility within the Pizzeria, is the moment he is hired by the Pizzeria and signs his employee contract. In this example we assume that the employee contract sufficiently describes all responsibilities of the employee in the Pizzeria. The part of the Construction Model of Pizzeria that shows the B-transactions and responsibilities needed to hire new personnel for Pizzeria is shown in Fig. 5 (N.B. this B-organization is in principle generic for the implementing of any enterprise with parties & people).

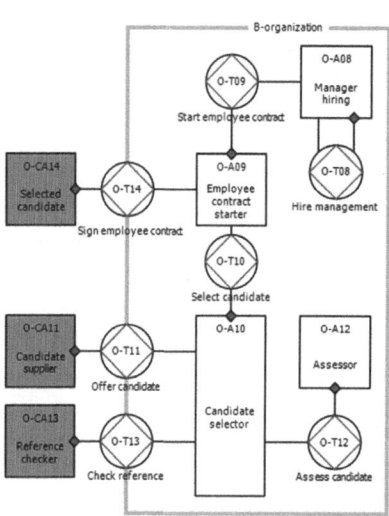

Fig. 5. Hiring personnel for Pizzeria

The model contains a B-actor role *Manager hiring* responsible for hiring new personnel. If new employees need to be hired in a certain Period, the Manager hiring will initiate one or more transactions *Start employee contract*. Before an employee contract can be started, the *Employee contract starter* needs a selected candidate to sign the contract. It is the responsibility of the *Candidate selector* to select a candidate. To be able to do this, the *Candidate selector* first requests the *Candidate supplier* to offer one or more candidates for the job. The Pizzeria uses an external party for this. Next, all candidates will be assessed by the *Assessor*. Based on the assessment the *Candidate selector* chooses a candidate. The last step of candidate selection is the checking of the references of the candidate (to make sure he or she is not part of the Italian mob that has a firm grip on the city) by an external party.

The management of the Pizzeria acknowledges that to enforce confidentiality of internally used information products, it must make the access to them depend on the authorization of an employee. Therefore from now on, each employee will receive a unique employee identity and specific employee authorization, based on the signed employee contract, not only to perform his business responsibilities, but also to have access to the information that is needed to fulfill them. The information products needed by the *Employee contract starter* are:

1. Share public identity information (source: selected candidate)
2. Share personal info (source: selected candidate)

3. Share selected candidate
4. Share employee information
5. Share job description

The *Employee contract starter* will first request public identity information from the selected candidate to be able to identify him as a specific person outside the enterprise (*Share public identity information*). The selected candidate must also provide some form of proof (e.g. a passport). With the provided public identity the necessary personal information of the selected candidate is requested (*Share personal information*). With the public identity the *Employee contract starter* will ask the I-organization to check if the selected candidate really doesn't already have an employee identity (*Share employee information*). If all is well, the *Employee contract starter* will create a new employee contract based on the job description of the job the new employee will fulfill. He will then request the selected candidate to sign the employee contract. After this is done the *Employee contract starter* creates a new employee identity, which is linked to the public identity of the new employee.

After the new contract is started, the *Employee contract starter* will request the I-organization to remember the newly created facts (*Remember start employee contract facts*). These facts consist of:

1. The employee identity
2. The public identity of the employee
3. The personal facts of the employee
4. The contract facts of the employee
5. The B-actor roles of the employee

From the last group of facts the necessary access to information products can be derived. The moment the *baker* tries to access an information product, the responsible I-actor role *Deriver order baking instructions* first checks the authorization of the *baker* by making use of the generic information product *Share employee authorization facts*. See Fig. 6.

Since this is generic, it is probably best to model this once in a separate generic diagram, to make sure the requirement to check the authorization of B-actor roles accessing information is not forgotten. Note that this confidentiality requirement can also be found in the top section of the Action Model, in every action rule specification, in the assess part, in the justice condition. This is the

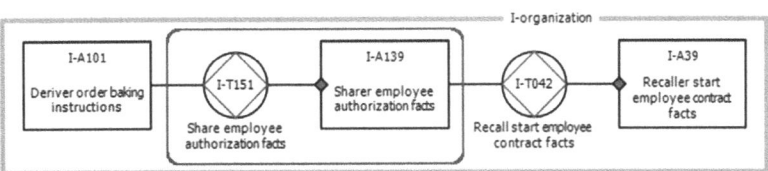

Fig. 6. I-organization supporting the *baker* extended with Sharing employee authorization facts

assessment whether or not the performer of the coordination act is identified and authorized to be the performer and whether or not the addressee of the coordination act is identified and authorized to be the addressee.

Conclusion. We conclude that in order to improve the confidentiality of the information product *Derive order baking instructions*:

1. Information about the identity and authorization of employees must be remembered in the I-organization.
2. Business processes must be in place that lead to the start, change and end of employee contracts on which the authorization of the employees are based.
3. The authorization of each B-actor role trying to access a confidential information product must first be checked by the I-actor role responsible for sharing the information product.

5   Artefact Demonstration and Evaluation

To demonstrate and evaluate our method, we applied its three steps to the real-life case of gathering requirements for a library information system for all public libraries in Flanders. We defined the information products that were created in the *Subscription* process, detected where they were used in the Library organization and focused on the use of these information products in the *Start loan* and *Establish library policy* processes. Interviews with different libraries led to a number of business requirements of which we used two in our thesis to determine relevant information quality characteristics. These business requirements were:

1. The library communicates *effectively* with its subscribers.
2. The library has a good insight in the geographical distribution of subscribers.

The second requirement proved a real concern during the interviews with some of the larger libraries who were trying to develop a special policy for subscribers coming from outside of the library city. The first requirement is important, among others, because with the introduction of a cashless library, a lot of subscribers postpone their payment for loans. When the amount to pay is higher than 10 euros, they receive an invoice from the library. Last year, 5% of the invoices returned because the address was incorrect. Effective communication in this case means invoices are sent to the right address. The number of addresses that are out of date therefore needs to be minimized. So, the nonfunctional business requirement of effective communication (QoB) leads to the nonfunctional information requirement (QoI) that subscriber addresses are *current*. This can be achieved in more than one way. We found four strategies to deal with currentness of address information. The first strategy requests the current address from the external source every time it is needed. The second strategy does it periodically. The third strategy only does it when the information is considered to be too old. The fourth strategy makes use of an update service that only

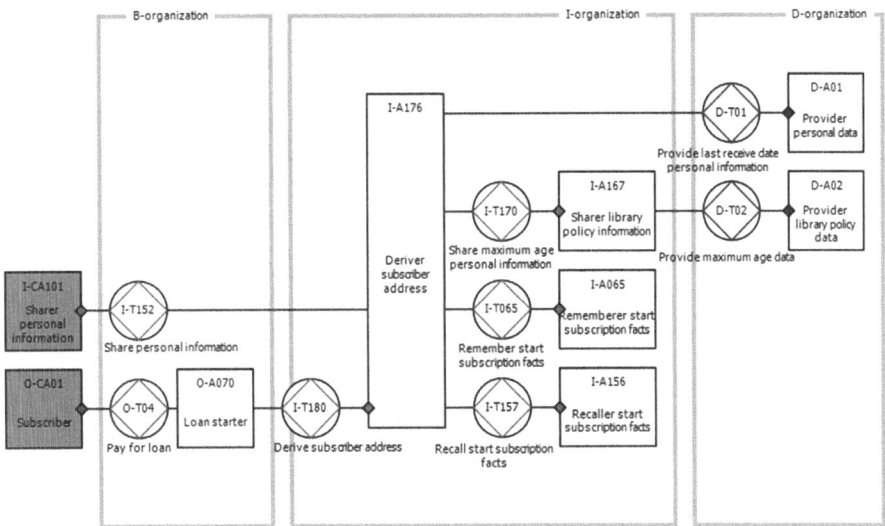

Fig. 7. Personal information is requested from the external source when the remembered facts are too old

sends information when there is an update of a certain address. As an example we will here only show the Construction Diagram of strategy 3 (Fig. 7).

In this strategy the way to deal with the need for current address information is to let the age of the subscriber address determine whether we use the information or refresh it by requesting the most current information from the external source. To achieve this we first need to define a new business transaction *Establish maximum age subscriber information* that may be periodically performed by Library Management. Second, we need to define a new I-actor role (*Deriver subscriber address*), who needs to check if the specific subscriber address that is requested by the B-organization has been received too long ago.

So, to find out if the subscriber address can be recalled internally or needs to be requested from the external source, the *Deriver subscriber address* will first request the maximum age of personal information, and second the last receive date of personal information (note this is a datalogical transaction!). By using this data the *Deriver subscriber address* can calculate if what needs to be done.

5.1 Conclusion

We conclude that to be able to improve currentness of *Derive subscriber address* for strategy 3 we have introduced:

1. a B-transaction to establish the maximum age of personal information
2. a D-transaction and corresponding D-actor role to provide the last receive date of personal information.
3. a new information product *Share maximum age personal information* and the corresponding I-actor role.

4. the I-actor role *Deriver subscriber address* to derive the subscriber address.
5. new action rules for the *Deriver subscriber address* to decide if the internally stored personal information is too old.

The benefits of applying our method in this real-life case were:

1. We were able to compare different strategies for achieving currentness of address information and make the tradeoff between the level of currentness and the costs of technical solutions to realize it.
2. The need for an I-actor role with the responsibility to derive a subscriber address with the selected level of currentness, led to specific requirements for the supporting library information system.
3. We were able to trace back implementation decisions regarding quality of information to business requirements.

6   Main Finding and Further Research

During to the development of our method, we came to the following main finding:

1. Applying our method leads to an ontological model that takes quality of information requirements into account, and is therefore more complete and more integrated with the I-organization.

We have shown a few examples of how the quality of information requirements led to the creation, remembering and recalling of new information, and of the use of this new information in the action rules of new or existing actor roles. In the Pizzeria case to take confidentiality into account we needed to model the ontology of hiring new personnel to be able to let I-actor roles check for the right authorization of B-actor roles requesting certain information.

In the real life case of the new library information system the need for effective communication (QoB) and current address information (QoI) led in strategy 3 to new information products, a new B-transaction, a new I-transaction and corresponding I-actor role, and a new D-transaction and corresponding D-actor role.

Further research could investigate if useful software requirements can be extracted from the enriched ontological models and in how far quality of information is *automatically* taken into account. Especially the new I-actor roles we found seem to be candidates for automation in a supporting information system. Other possibilities for further research are:

1. A further analysis of concepts like need, product and service. Though we defined some concepts necessary for our research, more investigation of terms like *information product* is needed.
2. Metrics of quality of information. We did not discuss the development of proper metrics for the quality characteristics of a particular information product. The quality characteristics of the ISO/IEC 25012 standard also have associated measurement methods and quality measure elements in the ISO

standard *Measurement of Data for Software product Quality Requirements and Evaluation* [8]. It would be interesting to investigate how to use this ISO standard to extend our method to also fit in the Measure phase of TDQM.

3. Metrics of quality of business. Since this is input for quality of information (cycle time, capacity of production/ coordination acts, etc.)

References

1. Deen, R.J.P., Mijs, J.: Adding quality of information requirements to the ontological model of an enterprise. Master thesis, Antwerp Management School (2016)
2. Adams, K.: Non-functional Requirements in Systems Analysis and Design. Springer, Cham (2015)
3. De Jong, J.: A Method for Enterprise Ontology based Design of Enterprise Information Systems. Dissertation, TU Delft (2013)
4. Dietz, J.: Enterprise Ontology. Theory and Methodology. Springer, Heidelberg (2006)
5. Dietz, J. (alias Perinforma, A.P.C.): The Essence of Organization. An Introduction to Enterprise Engineering, Sapio Enterprise Engineering (2013)
6. Dietz, J.: Architecture: Building Strategy into Design. Academic Service, The Hague (2008)
7. ISO/IEC 25012:2008: Software Engineering Software Product Quality Requirements and Evaluation (SQuaRE) Data Quality Model ·
8. ISO/IEC 25024:2015: Systems and Software Engineering Systems and Software Quality Requirements and Evaluation (SQuaRE) Measurement of Data Quality
9. Johannesson, P., Perjons, E.: An Introduction to Design Science. Springer, Cham (2014)
10. Hevner, A.R.: Design science in information system research. MIS Q. **28**, 75–105 (2004)
11. Krouwel, M., Op 'T Land, M.: Using enterprise ontology as a basis for requirements for cross-organizationally usable applications. In: MCIS (2012)
12. Van Lamsweerde, A.: Requirements Engineering: From System Goals to UML Models to Software Specifications. Wiley, Hoboken (2009)
13. Op 'T Land, M.: DEMO as core of Informed Governance at Rijkswaterstaat. In: EEWC (2013)
14. Op 'T Land, M., et al.: Enterprise Architecture: Creating Value by Informed Governance. Springer, Heidelberg (2009)
15. Op 'T Land, M., Proper, E.: Impact of principles on enterprise engineering. In: ECIS (2007)
16. Recker, J.: Scientific Research in Information Systems. A Beginners Guide. Springer, Heidelberg (2013)
17. Redman, T.: The impact of poor data quality on the typical enterprise. Commun. ACM **41**(2), 79–82 (1998)
18. Sommerville, I.: Software Engineering, 9th edn. Addison-Wesley, Boston (2011)
19. Wang, R.: A product perspective on total data quality management. Commun. ACM **41**(2), 58–65 (1998)
20. Whitten, J.L., Bentley, L.D.: Systems Analysis and Design Methods, 7th edn. McGraw-Hill, Inc., New York (2007)

DEMO/PSI Theory and the Law of the Land

Duarte Gouveia[✉] and David Aveiro[✉]

Madeira Interactive Technologies Institute, University of Madeira, Caminho da Penteada,
9020-105 Funchal, Portugal
duarte.gouveia@m-iti.org, daveiro@uma.pt

Abstract. This work analyzes two sources of law and elicit its underlying transactions. Then tries to model them using DEMO/PSI theory and analyzes the assumptions mismatches between law and DEMO/PSI. Design Engineering and Modeling for Organizations (DEMO) is a general-purpose theory and method to model interactions in society (between individuals and/or organizations) that uses a communication-centric approach. The Performance in Social Interactions (PSI) theory is a component of DEMO that explains a "universal transaction pattern" used to model those social interactions. The laws used as case studies are Portuguese contract law included in the Civil Code (from 1966); and the Common European Sales Law (CESL) from the European Union (EU), which is currently in final proposal stage. Through the analysis and discussion of these case studies, we suggest improvements to the DEMO/PSI theory based on the constraints expressed in those laws.

Keywords: DEMO · PSI theory · Common European Sales Law · Contract law

1 Introduction

Societies evolved to promote the "Rule of Law" over despotic government based on coercion from the strongest. The law sets rules, defines rights, duties and due procedures to promote trust, fairness and justice for all. But laws change over space and time. Each country has its own "Law of the Land", many times incompatible with what is established in the neighboring country, as things are valued differently.

This work uses two sources of law:

- Portuguese Contract Law is a section of Civil Code (Decreto-Lei 47344/66, 25/ November/1966 [2]) initially approved 51 years ago, but currently in its 69th revision [3]. We will address a section of this law (35 articles) that handles general conditions for contracts, content of contracts and time aspects of the process.
- Common European Sales Law (CESL) [4] is not yet an active law as it is still in the approval process by the European Union (EU) institutions. When the process started, in 2001, it aimed at harmonizing the European Contract Law. The current version is restricted to sales law and has been approved by the European Commission and Council in 2011 and has also been approved with recommendations by the European Parliament in a first reading in 2014. Although it is not yet an active law, from an academic perspective, the existing text provides relevant information that brings

© Springer International Publishing AG 2017
D. Aveiro et al. (Eds.): EEWC 2017, LNBIP 284, pp. 50–65, 2017.
DOI: 10.1007/978-3-319-57955-9_4

novel and useful knowledge regarding the transaction pattern. We will analyze the full extent of the law (186 articles).

CESL aims at being used as a second layer law that might be explicitly adopted by sellers and buyers on trades that cross-national borders, instead of the existing laws of both states.

Current law establishes that traders must fulfill the law of the land of the consumers [4]. This is a challenge for small and medium enterprises (SME) in EU with 28 different sets of laws to comply with. These constraints add transaction costs, that for SME mean a bigger share of company's turnover.

According to European Union (EU), only 10% of traders in EU export to other countries of the EU [4], and even those, only to a small number of countries. Among consumers in the EU, although 55% of consumers made online purchases, only 18% of consumers made purchases from a business in a different EU member state [6]. This raises concerns in the EU on the "differences in contract law between the EU member states that hinder traders and consumers who want to engage in cross-border trade within the internal market" [4].

Having a common law and a better model for sales might enable improved process awareness, flow, security and reduce transaction costs across borders.

Our hypothesis was: Can DEMO/PSI model the business processes established in the two sources of law identified above?

This paper is organized as follows. Section 2 reviews succinctly the DEMO/PSI theory [7]. Section 3 presents the research method used in this work. Sections 4 and 5 describe each of the two laws as case studies. Section 6 concludes summarizing the major contributions from this work.

2   Literature Review

The foundational theory of Organizational Engineering field is the Design and Engineering Methodology for Organizations (DEMO) [7]. A core idea of DEMO is that to model business interactions we should use a communication-centric approach, instead of the data-centric approach which is the dominant approach in the design of information systems.

The communication-centric approach has its roots in the Action Workflow Loop [8] presented in Fig. 1, being "general and universal", models the core pattern of all successful interactions.

Fig. 1. Action Workflow Loop [8]

According to Denning and Medina-Mora [9], "Incomplete workflows invariantly cause breakdowns, and if they persist, they give rise to complaints and bad feelings that interfere with the ultimate purpose of work – to satisfy the customer."

DEMO extends this core loop through Performance in Social Interactions Theory (PSI) [7, 10]. It describes the world through a model based on transactions, each producing a single result, initiated by a set of actor roles and executed by one particular actor role. This result is the simplified pattern presented in Figs. 4 and 5 which uses a sequence of coordination acts surrounding a production (execute) act.

As depicted in Figs. 2, 3 and 4, the transaction starts with a request (rq) by the initiator which includes the desired outcome in full detail. If the executor can fulfill that request, he will promise (pm) a delivery and then produce/execute the expected result and state (st) its completion to the initiator. Assuming that the delivered result is as requested, the initiator will finish the transaction by accepting (ac) the result, otherwise it can be rejected (rj). Therefore, this pattern assigns different acts to the initiator and the executor actor roles. These core acts can be split into three phases, as can be seen in Fig. 3: order, execution and result [11].

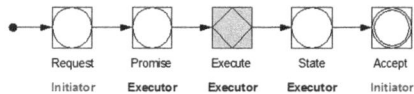

Fig. 2. Simplified pattern for a PSI transaction [7]

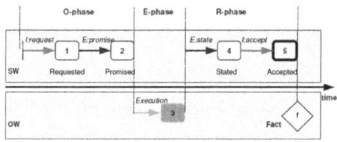

Fig. 3. Order, Execution and Result phases [11]

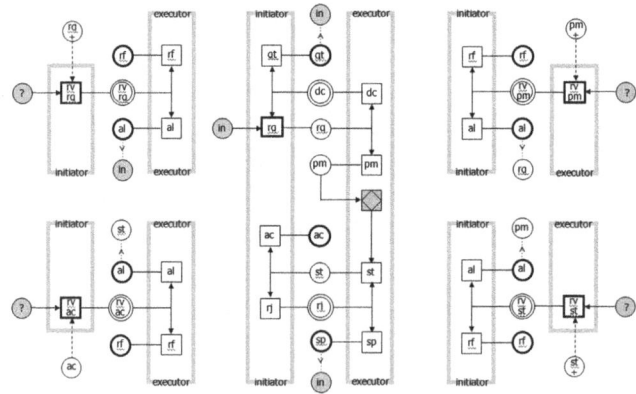

Fig. 4. DEMO 3.4 complete transaction pattern [12]

This simplified description becomes more complex, as can be seen in Fig. 4, as additional revoke acts are needed and so are added to each phase [10]:

- The *initiator* can change his mind and **revoke** the request (rv rq) at any time.
- The *executor* can **decline** (dc) the initial request if he does not wish, is not able, or can't deliver in the conditions requested by the initiator.
- The *executor* can **revoke** his previous **promise** act (rv pm).
- The *executor* can **revoke** his previous **state** act (rv st).
- The *initiator* may **reject** (rj) the **state**d (st) result.
- The *initiator* may **revoke** a previous **accept** (rv ac).

Revoking acts contradict previously established expectations. They may be initiated by any of parties and the counterparty may allow the revoke or refuse it. Please refer to the literature [7] for further information.

3   Research Design

To make research design options explicit, we adhered to the extension of this pattern presented in [13] and depicted the research process in Fig. 5.

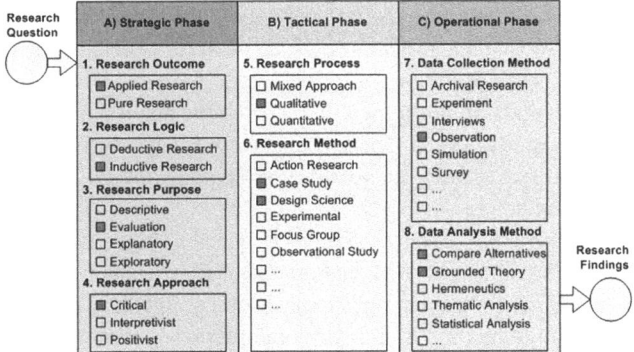

Fig. 5.   Research Design options using [13] pattern

This work is an applied research work (A1) that starts from a research question, presented in Sect. 1, and two case studies (B2) in a bottom-up general orientation.

From the Case Studies were analyzed using grounded theory method (C8) to detect more general concepts and patterns. Those results were also modeled using DEMO to develop model artifacts using Design Science [14] (B6). Both results were observed (C7) and compared (C8) with a critical approach (A4) and extracts its qualitative information (B5), which are the findings for this work. This work uses an inductive logic (A2) to evaluate (A3) DEMO/PSI assumptions and tools, confronting real world law examples with the current existing theoretical paradigm [15].

4 Portuguese Civil Law

4.1 Definitions in Portuguese Contract Law

What is stated by the Law

Portuguese Civil Law [2] defines an "obligation" (art. 397) as a lawful bound that a person has with another, to perform a certain act. Obligations can be enforced by the states through coercion, with one exception. A "natural obligation" (art. 402, 404) is an act, founded in a duty to justice, that is not mandatory by law.

A "contract" is an agreement between the parties, which is a special case of obligations. A "unilateral promise" is a "promise contract" where only one of the parties has established an obligation. A "reciprocal contract" (or "bilateral contract") is a "contract" where both parties have established obligations to one another.

In a "promise contract" parties create obligations to establish a future "contract" with certain terms (art. 410 .1), as they are not presently able to fulfil them, typically because depend on acts from third parties. The promise contracts are common on sales that require registry, certificates, licences or other requirements determined by law (art 410 . 3). Promise contracts can be transmitted to the heirs, in case of death of the initial parties (art. 412).

Modeling it with DEMO/PSI

DEMO/PSI assumes participants always act freely in transactions [7]. The literature does not reference how to force an act by coercion by a court decision, neither its effects, which might be complex and hard to fully predict.

Obligations, as defined by Portuguese law, match the result phase of the transaction (state-accept), as depicted in Fig. 2. The agreement established by contract creates those obligations. The contract corresponds to the order phase of DEMO/PSI transaction (request-promise), as depicted in Fig. 2.

DEMO/PSI assumes that each transaction produces a single transaction result. The law does not force that constraint. On the contrary it assumes the opposite by defining reciprocal contracts. The contract may include several obligations from one party to the other. Modelling each obligation as a distinct transaction ignores the fact that they are approved together (order phase).

The Portuguese law clearly distinguishes "unilateral promise" from "reciprocal contract". This not only contradicts the single result assumption, but also the assumption that establishes fixed roles, assigning to the initiator actor role the acts of requesting and accepting, and to the executor actor role the acts of promising, executing and stating. This topic has been discussed in the literature [16–19].

This conceptual mismatch is clarified by distinguishing the contract (order phase), from the obligations (result phase). DEMO/PSI transactions bundle both phases into a single transaction. We believe a good improvement to DEMO/PSI theory would be to split the two phases. The contract is the mutual agreement that participants reach after discussing the expected results. Contracts establish intents and promises for the future - one or more obligations - not its concrete execution.

Each obligation should be a transaction with one executor, one beneficiary and one result. When executed, stated and accepted, the transaction result fact is enacted.

Natural obligations are an interesting case, as they do not emerge from a contract. A natural obligation is optional, and lives on the discretionary decision scope of its executor. This special case provides additional reasons to the splitting of DEMO/PSI transactions into the order phase and the result phase.

4.2 What Is the Content of a Contract?

What is stated by the Law

Parties can freely agree on contracts (art. 398 .1, 405 .1), and establish as many terms as they wish, even including rules from other contracts (art. 405 .2).

The contract might not include money, but must be something that the receiver values and is worthy of legal protection (art. 398 .2). The contract contents (art. 398 .1) can be a positive (things to do) or a negative (things to abstain from doing).

The content of the contract can be undetermined when agreed upon, because the precise terms might be trusted to any of the parties or to a third party, using fairness as ruling principle (art. 400 .1). In case of disagreement, the decision is up to a court of law to clarify the terms of the contract (art. 400 .2).

The transfer of ownership is determined by contract, except in the exceptions determined by law (art 408 .1). In general, ownership is transferred when the thing is acquired by the seller (art 408 .2). In case of undetermined things, the ownership is transferred when the thing is determined and both parties acknowledge it. In case of natural fruits, component parts or parts of a whole, ownership is only transferred, by default, at the moment of separation (art. 408 .2). In case of transfer of ownership for goods that require registry, like properties, registry date is the reference (art 409 .2).

The seller might reserve the ownership rights until total or partial fulfilment of the obligations established by the counterparty or until any established event (art 409 .1).

Modeling it with DEMO/PSI

Portuguese law allows the free establishment of contracts. The interdependencies that parties may agree upon in contracts may be difficult to model in DEMO/PSI.

Among the things that the receiver might agree upon is the "negative content", that is, the absence of performing a certain act, or not to do it during a certain period. In general, facts in DEMO/PSI are positive facts. This is a novel issue that must be noticed by the academic community.

A negative act is a recognition that an act did not occur. Unless an act is public, only one of the parties can do that recognition.

The Portuguese law allows contracts where the content is not fully determined. For example, the next year fruits from apple trees in an orchard might be sold without knowing quantity, quality or price by which they will be sold. They will need information acts to establish that additional information, which may be subject to disagreement and dispute. This contradicts current DEMO/PSI assumptions where all details of a transaction are fully established when a request is performed.

The Portuguese law introduces a conceptual distinction between the actual transfer of the goods and the ownership of the goods. The location of the goods does not determine ownership. Ownership is an inter-social concept that the law regulates. Some cultures don't even have the concept of ownership. Things might just belong to everyone or to nature, or to the country. This topic is further developed in Sect. 5.3.

4.3 Who Is Who in Contracts

What is stated by the Law

If, by successive contracts, different people gain rights over the same thing that are incompatible, it takes precedence the oldest contract (art. 407).

In "reciprocal contracts", any of the parties may delegate his position to a third party, as long as counterparty agrees on that transfer, either at start or during the execution of the contract (art 424 .1). If agreed at start-up it only becomes active with the acknowledge of the delegation (art 424 .2).

A "preference pact" is an agreement where a party assumes the obligation of giving preference to someone in the transfer of ownership of a thing (art. 414). Preference pacts are not able to be transferable to others, either in life or death (art. 420). The law establishes the duty to inform of the intention to sell, the default time to decide on that preference, the right to bundle things that significantly increase value if sold together, how to act when several entities have the right to preference, and the situations that cancel the right to preference (art. 416 to art. 423).

Modeling it with DEMO/PSI

DEMO/PSI establishes that the assignment of persons to transaction actor roles is performed by someone with the authority and responsibility. A person in an assigned position cannot freely delegate its role assignment to a third person, but this law allows it if there is agreement by the counterparty in the transaction.

It is also not sufficiently clear in DEMO/PSI the distinction between persons and legal persons. Not all persons are legal persons, due to under age, incapacity or illness issues. But organizations might be collective legal persons. This is a topic that requires additional clarification in DEMO/PSI theory.

This law says that delegation become active only after the counterparty as acknowledges the information act. The assumption that the mere delivery (through e-mail, physical delivery or digital notification on a channel or device) is enough to acknowledge is clearly disputable and a reason for substantial disagreement and abuse in the social world.

The preference pact rule elicits the need of question acts or timed events with default answer. A counterparty decides on a right they have. We detected a general pattern for enacting rights from what occurs in preference pacts, which is different from the execution of obligations. Obligations are expected to occur according to the PSI pattern, in particular the result phase as stated in Sect. 4.1. Enacting rights act are optional. When they occur, they change the inter-social context by mere notification, not by agreement. The enacting of the right might be contested by the other party, leading to the need to an agreement or an external ruling. Enacting rights enable or disable other acts as designed in each business process.

4.4   How Contracts Are Executed?

What is stated by the Law

A contract can establish initial conditions, suspension conditions, conditional terms, and until conditions, with validity time periods (art. 401 .2).

The impossibility of performing an act makes the contract null – as if it never existed (art. 401 .1), but only when that constraint is from the object, not from the person with an obligation (art. 401 .3).

The contract should be fulfilled by parties as initially agreed upon. The termination of the contract or any change to it requires mutual consent (art. 406 1.).

Modeling it with DEMO/PSI

DEMO/PSI transaction pattern defines initial conditions and revoking conditions through agreement. Suspension conditions hold the execution of the contract, that is, the obligations not yet performed, but do not destroy the initial agreement. Suspension conditions can be enacted by an information act, that might be disputed by the counterparty. These conditions can have time periods that enact changes in the inter-social state. These requirements are not currently handled by DEMO/PSI.

Conditional terms and Until conditions might terminate the contract. We might argue that these terms can be modelled in DEMO action rules, but action rules only enable or prevent existing acts to be performed, and do not allow new coordination acts to take place. _These terms follow the right pattern previously elicited. Further discussion on this topic on Sect. 5.5.

The Portuguese law establishes that the contracts must be fulfilled as initially agreed upon. DEMO/PSI is more flexible and allows the initiator to accept the production result through a state act, even if it does not comply with the initial agreement. We believe the Portuguese law should accommodate this option, as it enlarges the options to the beneficiary. DEMO/PSI also states that even if no state result is produced, the initiator should be able to accept. This however is not supported by the current PSI transaction pattern, as defined in [12]. Section 5.1 will provide additional clarification on the discussed topics in this section.

4.5   When – Temporal Aspects of Contracts

What is stated by the Law

A contract can establish future dates for the fulfilment of its terms (art. 399). If a contract that does not establish milestones for obligations by the parties, each party can refuse to fulfil its duties until the counterparty fulfils theirs, or do it simultaneously (art. 428 .1), or provide additional guarantees (art. 428 .2). If there are changes in the circumstances that diminish guarantees by a party, the other party may refuse to fulfil its acts (art. 429, art. 430). This situation may be reverted if a party delegate their rights and obligations to a third party (art. 431).

If there was no time limit established initially, the person in with an obligation may request the setting of a time limit, after which the obligation ends (art. 411).

Modeling it with DEMO/PSI

The contract may include timely defined events that change the inter-social world. DEMO/PSI describes a discrete event system where at each moment in time a set of events can take place in unpredictable sequence. Each event might induce a change in the social-world state and therefore on the possible events that become available.

DEMO/PSI does not allows events that change the inter-social world without the deliberate act of parties. That is clearly a limitation of current theory. If terms have been established in the contract based on the passage of time, when due time has passed, the inter-social world should change accordingly. It should be assumed to be implicit acts that was previously agreed upon in the contract.

Portuguese law has a unclear and confusing formulation regarding the "changes in the circumstances that diminish guarantees by a party", and its consequences, as these circumstances are subjective by each party.

5 European Common Sales Law (Proposal)

5.1 Reaching a Contract – the "Contract Resolution"

What is stated by the Law

The ECSL aims at regulating sales contracts, including digital content and related service. Contracts are agreements that give rise to obligations between parties (art. 2), as they can freely agree on the terms of the contract (art. 1). The ECSL distinguishes the cases where the buyer is a consumer or a trader (art. 23).

Communication between the parties is performed with "notices" (art. 10). A notice becomes effective when it reaches the addressee, unless it provides for a delay effect. A notice has no effect if a revocation of it reaches the addressee before the notice.

The sales process starts with an "offer" presented by the seller (art. 31), which is the most common case, or by the buyer (art. 38). Offers from the seller promoting its products and services can be done to the public (or to some segment) or directly (to a particular person or organization). If the offer is public, it must be requested by the buyer and accepted by the seller. If the offer is directed to a buyer, then it can be just explicitly accepted (art. 34, 35) or rejected by default (art. 30, 33).

The offer might have a time limit for acceptance (art. 32, 35, 36, 39). The ECSL establishes that in late acceptances (art. 37) it is up to the seller to explicitly clarify its option to confirm it or not.

A buyer might notify a seller with a modified acceptance (art. 38). According to the ECSL (art. 38), a modified acceptance should be considered as a rejection of the initial offer, and the creation of a new offer. That new offer is initiated by the buyer, with the terms of the original offer, changed by the modified terms. The seller can then decide if he accepts the offer or not. If the seller responds with a modified acceptance, then again it should be considered as a rejection of the offer and the creation of a new one, this time initiated by the seller with the new bundle of terms.

The offer may be revoked by the party that proposed it at any time (art. 32). However, if the counterparty has already "sent an acceptance" or in cases of "acceptance by conduct", the contract may be assumed to be established (art. 32, 35). The ECSL introduces ambiguity with these ruling.

The ECSL distinguishes the terms of the contract that were subject to previous discussion or individual negotiation (art. 7) and those that were proposed by the seller and accepted by the buyer without individual negotiation of the terms. A term is said to be individually negotiation if the buyer has the ability to influence the content of the terms. When the buyer is a consumer, the burden of proof on individually negotiated terms is on the seller, even if the terms of the contract were written by a third party. When the buyer is a trader the burden of proof on individually negotiated, terms is on the party making that claim.

When a contract is established it is said to be concluded (art. 30, 35), but not in a very clear way, in the authors opinion. Contract conclusion should not to be confused with contract termination (art. 8) that are handled at Sect. 5.5.

Modeling it with DEMO/PSI

Just like the Portuguese law, the ECSL defines a contract as something that can have with multiple fulfilments by multiple parties.

The ECSL clearly states that communication between parties is made through notifications and that each notification only becomes effective when acknowledge by the counterparty, except in a delay effect is included in the message. As current electronic communications are so fast, we believe there should be a mandatory quarantine period so that the sender can detect a mistake and correct it. It could also be a beneficial situation for the receiver as revoking acts because of cancelations can be a costly process and prone to errors.

The ECSL states that the business transaction starts with an offer. This is aligned with ontological constructions for services [20], but not with DEMO/PSI [7] that assumes it to be initiated by the transaction beneficiary with a request.

ECSL describes the negotiation phase in a turn based interaction with detailed instructions on the importance of individually discussed terms. The discussion of the terms between parties is not yet a common practice in electronic commerce, but it is an idea that is creating its path in law scholars [21]. The ECSL tries to reduce the intervention of courts of law (or equivalent arbitrary decisions) to decide on things parts can freely agree upon. Portuguese law call for action from courts more often.

DEMO/PSI does not prescribe how the negotiation takes place. ECSL assumes a rigid turn based process. In the author's opinion the actual negotiation process isn't that important, as long it is clear what terms are included in the agreement, which terms were individually discussed, that all important issues are covered and that a clear agreement was reached and acknowledged.

5.2 Withdrawal from Contract and Defects in Consent

What is stated by the Law

For contracts established at distance or off-premises between traders and consumers, the ECSL allows the consumer to withdraw from a contract within 14 days after it was concluded, without any need to justify that position (art. 40). There are exceptions to this right (art. 40), as it cannot be applied to food and beverages or others liable to deteriorate or expire rapidly like newspapers or magazines, goods or digital content that were made to the customer specification or clearly personalized or that has started to be provided without a tangible medium, or other sealed goods.

The withdrawal is established by simple notification and does not require agreement by the seller. The seller must acknowledge such a withdrawal without a delay. (art. 41). The ECSL establishes a complex rule for determining when those 14 days start to count (art. 42).

In the case of withdrawal, buyer and seller are no longer obliged to the terms of the contract, but still have new obligations to fulfil to the counterparty (art. 17, 44, 45) regarding restitution, as presented in Sect. 5.5.

The ECSL also distinguishes several forms of defects in consent, namely: mistakes; fraud; threats and unfair exploitation:

- Mistake of fact or law (art. 48), when: (a) a party would not have concluded the contract or would have done so only on fundamentally different contract terms; (b) the other party knew or could be expected to have known this; (c) the other party caused the mistake or did the same mistake, or failed to inform the counterparty as expected based on pre-contractual information duties.
- Fraud (art. 49), if the other party has induced the conclusion of the contract by fraudulent misrepresentation, whether by words or conduct, or fraudulent non-disclosure of any information which good faith and fair dealing required that party to disclose, especially if the other party had special expertise or information.
- Threats (art. 50), if the other party has induced the conclusion of the contract by the threat of wrongful, imminent and serious harm or of a wrongful act.
- Unfair exploitation (art. 51), if at the time of the conclusion of the contract a party was in economic distress, had urgent needs, was improvident, ignorant or inexperienced and the other party knew or could be expected to have known this and, in the light of the circumstances and purpose of the contract, exploited the first party's situation by taking an excessive benefit or unfair advantage.

An avoidance based on the previous terms, is established through notice to the counterparty (art. 52). The avoiding party has a period to avoid after becoming aware of the relevant circumstances or becomes capable of acting freely. That period is six months in case of mistake and one year in case of fraud, threats and unfair exploitation. If the party who has the right to avoid the contract, has become aware of the circumstances that might lead to avoid, but confirms the contract, either expressly or impliedly, it may no longer avoid the contract (art. 53).

The effects of avoidance (art. 54) can be limited to only certain contract terms. A contract which may be avoided is valid until avoided but, once avoided, is retrospectively invalid from the beginning. Rules of restitution, as presented in Sect. 5.5, may be applied to an avoided contract. A party may have the right for damages for loss, provided that the other party knew or could be expected to have known of the relevant circumstances (art. 55).

Modeling it with DEMO/PSI

Withdrawal is a novel coordination act that doesn't exist in DEMO/PSI. It is an unilateral act (right) with a limited time bound to be enacted, that requires acknowledge by the counterparty and has the same effect as a allowed revoke request in DEMO/PSI.

Defects in consent is a very interesting part of the ECSL. DEMO/PSI assumes that in each coordination act there is always a claim to truth, to justice and to sincerity, otherwise the coordination act does not take place. The ECSL does not assume any of those constraints, but allows an act to be avoided for extended periods.

We could argue that the defects in consent can be modelled by existing revoke request or revoke promise acts in DEMO/PSI. The main difference is that on revoke acts, they only take effect with agreement by the counterparty, while in ECSL they just have to be acknowledge to become effective, even if later on contested by the counterparty. It is the contestation that requires agreement, not the avoidance. The avoidance takes immediate effects.

5.3   Obligations and Rights

What is stated by the Law

The ECSL distinguishes: (a) the transfer of the goods or fulfilment of services; (b) the transfer of ownership; (c) the acceptance of goods (or services) after checking conformity; (d) the transfer of risk; (e) the installation of goods; (f) accepting the installation of goods (art. 91, 140–146).

The seller has responsibilities regarding the carriage of the goods (art. 96). In the case of transfer of digital content, the initiative of transferring of the content might be of the buyer (art. 97).

In the case of delivery of physical goods the ECSL establishes the terms in which the delivery will occur, namely place (art. 93), method (art. 94) and time (art. 95). A delivery to be performed earlier than expected is subject to rules. Delivery of goods might require installation, maintenance and repair obligations to be performed by the seller (or a third party he delegates it to). If not agreed otherwise, it is expected that the delivery takes place within 30 days (art. 15) from the date of conclusion of the contract. The ECSL treats differently lack of conformity in the goods and in the installation of the goods. The contract might also require maintenance obligations. If the performance of those obligations requires seller access (or third parties he delegates to) to the buyer's facilities, it is subject to previous scheduling and approval.

A contract made up of divisible components can be partially fulfilled if only some of the obligations are fulfilled. A party may withhold some obligations if the counterparty has not fulfilled theirs in due time (art. 113). The law establishes remedies to solve unfulfilled parts of the contract, as presented in Sect. 5.4. It is possible to terminate only parts of the contract. The ECSL distinguishes between fundamental and non-fundamental non-performance (art. 87).

If the buyer is a trader, he might request the examination of the goods (art. 121, 122) in the delivery chain. Lack of conformity must be notified to the counterparty.

Both parties can delegate the performance of its obligations or transferring its benefits to a third party (art. 78, 92), unless the contract explicitly requires the "personal performance" by the seller.

Both buyer and seller can, in anticipation, inform the counterparty of a temporal constraint that prevents them from performing an obligation in due time. Those acts are called "excused non-performance" (art. 88). The excused non-performance does not require agreement by the counterparty. The counterparty can establish a reasonable time limit to overcome that limitation, that if not fulfilled in time, becomes a reason for termination. The non-performer can contest the proposed reasonable time when that time constraint is established, but not later on (art. 88).

The ECSL references "changes of circumstances" (art. 89), but does not defines bounds to it. Even if the cost of performance has increased its value or what is to be received has diminished, that is not a reason to force re-negotiations on an established contract. Only on "exceptional circumstances" parties have a duty to negotiate, adapt or terminated the contract.

The ECSL establishes the terms for payments (art. 123–128) as one of the obligations that have to be fulfilled by the parts. Different circumstances can give rise to obligations to pay from both parties, as defined in Sect. 5.4.

Modeling it with DEMO/PSI

The separation of several transfers within a single sale contract provides additional evidence to the need to separate the order-phase from the result-phase in DEMO/PSI.

The ECSL introduces a notion of excused non-performance, through mere notification. That situation, constrains the rights of the counterparty regarding the termination of contract by non-performance, but creates a right to set a time limit. DEMO/PSI does not provide for any mechanism to enforce this in the standard transaction pattern. Promised times in DEMO/PSI are merely indicative, but not enforced in any way.

Regarding delegation, the ECSL provides a different legal method than what was presented on Sect. 4.3. The Portuguese law accepts delegation at any time if the counterparty accept it. The ECSL allows delegation at any time through mere notification, unless the contract requires the personal performance. The delegation can occur either on the provider or on the beneficiary sides of the transaction. The authors believe that changing the legal persons in the transaction is a substantial change that should not occur unless the counterparty agrees to, either initially in contract terms or through acceptance during execution.

5.4 Remedies, Damages and Interest

What is stated by the Law

The ECSL establishes remedies (art. 106–112, 147–158) for damages, delays, costs they might produce, non-performance and performance below what was reasonable to expect from the contract. Delays in payments (art. 159–171) are also subject to interest at an establish reference rate (EU Central Bank or National Banks) plus 2%.

Modeling it with DEMO/PSI

DEMO/PSI does not establish any rulings on how to get out of discussion states, except that it has to be done through agreement between the parties.

The ECSL follows the same principles, but lists the possible remedies. All remedies can be modelled by parties agreeing upon a set of new obligations. This provides additional evidences to the idea that a transaction should be split between ordering and result phase, and that the result phase should have multiple obligation fulfilment transactions, each producing its own transaction result.

5.5 Termination and Restitution

What is stated by the Law

Terminating a contract (art. 8) means to bring to an end the rights and obligations of the parties under the contract, except of those that are explicitly applicable even after termination, like settlement of disputes, payments due and damages for non-performance established before the time of termination.

The reasons that might lead to a termination are the non-performance of obligations (art. 87, 114–116). An obligation is not performed if there wasn't the timely delivery of the goods or digital content in conformity with the contract, or payments were not executed or were executed later than expected.

ECSL establishes (art. 88) that a fundamental non-performance is one that "substantially deprives the other party of what that party was entitled to expect under the contract" or that anticipates non-performance as the party cannot be relied upon.

A contract that was established with undetermined duration can be terminated with a reasonable period of notice, not exceeding two months (art. 77).

When a contract is avoided or terminated there might emerge the rights of restitution (art. 17, 44, 45, 172–177). Namely the return of the goods and the full amounts made available by the counterparty within 14 days from the withdrawal notice. The seller may withhold the reimbursement until it has received the goods back. The consumer must bear the costs of returning the goods and is responsible for the diminished value of the goods if improperly handled.

Modeling it with DEMO/PSI

The conditions established in ECSL for terminating and agreement, in the authors opinion, are equivalent to reaching a change agreement, that is, replacing the existing obligations with new ones, even if the new are restitution obligations.

In the authors opinion, the only relevant change in a transaction that justifies having a special inter-social state that marks the contract as terminated is when there is no additional obligation to fulfil and for that moment on no new obligation can be created through the enacting of any right. In DEMO/PSI terms it would mean no additional coordination acts are to be performed, including revoke acts.

6 Conclusions

This Section summarizes the findings presented in Sects. 4 and 5:

- There are at least six types of acts that change the inter-social state of the world:
    (a) Coordination acts freely taken by actors according to the transaction pattern;
    (b) Negative-acts, the absence of some party to perform a certain act;
    (c) Information acts that notice changes in context (by parties or third parties);
    (d) Question acts that require an answer, eventually associated with a default answers and a time limit;
    (e) Enacting of Rights by mere declaration, taking immediate effect, but that might be disputed later on, reverting its effect;
    (f) Previously agreed timed based acts enacted by agreed time or elapsed time;

(g) Acts enforced by a court or by death of a party;
- Contracts are always established by legal persons (either capable persons or recognized legal organizations).
- DEMO/PSI transactions could be split into distinct transactions for the Ordering Phase and the Result Phase, with benefits in modelling business transactions.
  (a) The ordering phase is an agreement that expresses the intention of the parties to perform a certain bundle of obligations (one or more).
  (b) The ordering phase might be established in an incomplete form, requiring later notifications or agreements on the missing terms.
  (c) The result phase corresponds to the fulfilment of a specific obligation by one party in the benefit of another. The agreement may include several obligations, performed by different parties and to the benefit of others.
- All coordination acks require acknowledge by the counterparty, or a subsequent act that tacitly acknowledges it.
- It should be possible for a party at the top to delegate role and responsibilities beneath him to a third party with agreement by the counterparty.
- The conclusion of agreements requires a final acknowledge that makes it active.

All models simplify reality and therefore are never complete. Models should be judged by their usefulness, not completeness. The current DEMO/PSI transaction pattern is not able to model the complexity of the laws used as case studies. Some minor changes to DEMO/PSI, as the ones suggested on this work, could largely increase conformity with the law, without diminishing DEMO/PSI current use. We believe the suggestions presented above are general, but that must be subject to validation on future studies.

Acknowledgments. This work was partially funded by FCT/MCTES LARSyS (UID/EEA/50009/2013 (2015-2017)).

This work was developed with financial support from ARDITI (Agência Regional para o Desenvolvimento da Investigação, Tecnologia e Inovação), in the context of project M14-2009–5369-FSE-000001-Bolsa de Doutoramento.

References

1. http://www.dgpj.mj.pt/sections/leis-da-justica/pdf-ult/sections/leis-da-justica/pdf-ult/codi-comercial-de-1888/downloadFile/file/CodComercial.pdf?nocache=1188821262.8
2. https://dre.pt/application/dir/pdf1sdip/1966/11/27400/18832086.pdf
3. http://www.pgdlisboa.pt/leis/lei_mostra_articulado.php?nid=775&tabela=leis&so_miolo=
4. http://eur-lex.europa.eu/legal-content/EN/TXT/?uri=celex:52011PC0635
5. http://www.europarl.europa.eu/oeil/popups/summary.do?id=1339866&t=e&l=en
6. https://www.out-law.com/en/articles/2015/june/common-european-sales-law-proposals-to-be-replaced-as-new-consultation-is-opened-on-online-sales-barriers/
7. Dietz, J.L.G.: Enterprise Ontology – Theory and Methodology (2006)

8. Medina-Mora, R., Winograd, T., Flores, R., Flores, F.: The action workflow approach to workflow management technology. In: Proceedings of the 1992 ACM Conference on Computer-supported Cooperative Work, pp. 281–288. ACM, December 1992

9. Denning, P.J., Medina-Mora, R.: Completing the loops. Interfaces **25**(3), 42–57 (1995)

10. Dietz, J.L.G.: DEMO-3 Way of Working (2009)

11. Van Reijswoud, V.E., Mulder, H.B., Dietz, J.L.: Communicative action-based business process and information systems modelling with DEMO. Inform. Syst. J. **9**(2), 117–138 (1999)

12. Dietz, J.L.G.: The PSI theory – understanding human collaboration (v3.2) (2016)

13. Wohlin, C., Aurum, A.: Towards a decision-making structure for selecting a research design in empirical software engineering. Empir. Softw. Eng. **20**(6), 1427–1455 (2015)

14. Gregor, S., Hevner, A.R.: Positioning and presenting design science research for maximum impact. MIS Q. **37**(2), 337–355 (2013)

15. Kuhn, T.S.: The Structure of Scientific Revolutions. University of Chicago Press, Chicago (2012)

16. Rittgen, P.: Negotiating models. In: Krogstie, J., Opdahl, A., Sindre, G. (eds.) CAiSE 2007. LNCS, vol. 4495, pp. 561–573. Springer, Heidelberg (2007). doi: 10.1007/978-3-540-72988-4_39

17. Weigand, H., De Moor, A.: A framework for the normative analysis of workflow loops. ACM Siggroup Bull. **22**(2), 38–40 (2001)

18. Lind, M., Goldkuhl, G.: Questioning two-role models or who bakes the pizza. In: Seventh International Workshop on the Language-Action Perspective on Communication Modeling (LAP 2002), p. 44, June 2002

19. Dietz, J.L.: Isn't baking a pizza that easy? In: Seventh International Workshop on the Language-Action Perspective on Communication Modeling (LAP 2002), p. 65, June 2002

20. Quirino, Glaice K., et al.: Towards a service ontology pattern language. In: Johannesson, P., Lee, M.L., Liddle, Stephen W., Opdahl, Andreas L., López, Ó.P. (eds.) ER 2015. LNCS, vol. 9381, pp. 187–195. Springer, Cham (2015). doi:10.1007/978-3-319-25264-3_14

21. de Rosnay, M.D.: Peer-to-peer as a design principle for law: distribute the law. J. Peer Prod. **6**, 1–9 (2015)

The Perspectives of DEMO Application to COSO Internal Audit Framework Risks Mitigation

Eduard Babkin[1], Pavel Malyzhenkov[1(✉)], and Fabrizio Rossi[2]

[1] Department of Information Systems and Technologies,
National Research University Higher School of Economics,
Bol. Pecherskaya 25, 603155 Nizhny Novgorod, Russia
{eababkin,pmalyzhenkov}@hse.ru
[2] Department of Economics and Enterprise, University of Tuscia,
Via del Paradiso, 47, 01100 Viterbo, Italy
fabrizio.rossi@unitus.it

Abstract. The Committee of Sponsoring Organizations of the Treadway Commission (COSO) Framework is one of the most diffused internal audit (IA) tools. According to some sources 82% of companies use it realizing the IA procedures as their reference framework. Still, it presents some limitations, especially in the field of risk assessment linked to the weak formal distinction between the framework itself (the organizational structures and policies realized to promote, integrate and improve the management of risk) and the process used for risk management (the operations realized to assess, treat and monitor the risk). So, the formal separation of the organizational level from the aspects which the framework seeks to integrate into all critical organizations processes where decisions are made is needed. The COSO framework was realized on the base of principles-based approach. So, the aim of the present paper is to describe the formal approach based on DEMO methodology tools oriented to the elimination or mitigation of limits inherent to COSO framework.

Keywords: Internal audit · COSO framework · DEMO · Interaction model

1 Introduction

Committee of Sponsoring Organizations of the Treadway Commission [4] defines internal control as ''a process, effected by an entity's board of directors, management and other personnel, designed to provide reasonable assurance regarding the achievement of objectives in the following categories:

- effectiveness and efficiency of operations;
- reliability of financial reporting;
- compliance with applicable laws and regulations''.

This definition reflects certain fundamental concepts. Internal control is:

- geared to the achievement of objectives in one or more categories—operations, reporting, and compliance;

© Springer International Publishing AG 2017
D. Aveiro et al. (Eds.): EEWC 2017, LNBIP 284, pp. 66–73, 2017.
DOI: 10.1007/978-3-319-57955-9_5

- a process consisting of ongoing tasks and activities—a means to an end, not an end in itself;
- effected by people - not merely about policy and procedure manuals, systems, and forms, but about people and the actions they take at every level of an organization to affect internal control;
- able to provide reasonable assurance - but not absolute assurance, to an entity's senior management and board of directors;
- adaptable to the entity structure - flexible in application for the entire entity or for a particular division, operating unit, or business process.

This definition is intentionally broad. It captures important concepts that are fundamental to how organizations design, implement, and conduct internal control, providing a basis for application across organizations that operate in different entity structures, industries, and geographic regions. Actually, the idea of a principles-based approach to standard setting is not new. The Board's conceptual framework contains the body of principles that underlies business activity conducting principles. The Board has used the conceptual framework in developing the principles for more than 20 years. However, many assert that the standards have become increasingly detailed and rules-based (with "bright-lines" and "on-off" switches that focus on the form rather than the substance of transactions), complex, and difficult and costly to apply. Many also assert that the standards allow business engineering to structure transactions "around" the rules, referring to situations such as those in which complex structures or a series of transactions are created to achieve desired results (Fig. 1).

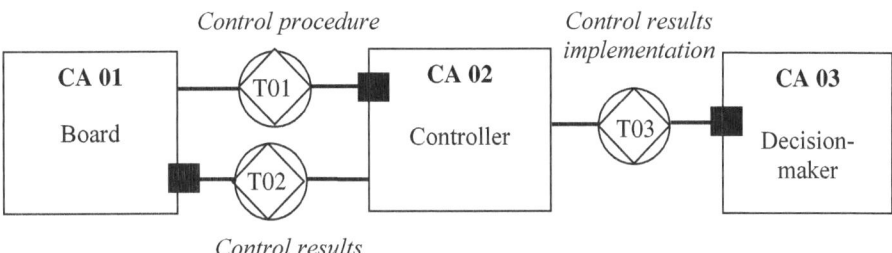

Fig. 1. Actor–transaction diagram

Under a principles-based approach [1, 3, 4], the principles in audit and control standards would continue to be developed from the conceptual framework, but would apply more broadly than under existing standards, thereby providing few exceptions to the principles. In addition, different standard-setting bodies would provide less interpretive and implementation guidance for applying the standards. Because exceptions and interpretive and implementation guidance are largely demand-driven, a principles-based approach would require changes in the processes and behaviors not just of different standard-setting bodies, but of all participants in the accounting and reporting process including preparers, auditors, control bodies like SEC or CONSOB (Italian Securities and Exchange Commission) and users of financial information.

All these risks present in Internal Audit field can be eliminated by using formal reference models constructed on the DEMO base as a tool which offers the possibilities of formal approach to audit process engineering. Besides, DEMO may contribute in the mitigation of COSO risks linked to a weak distinction of responsibilities between the policy-setters and policy-implementators. So, the idea is that applying the DEMO apparatus which stresses much on authority and responsibility concepts this disadvantage can be overcome. The paper is organized as follows: Sect. 2 analyzes the COSO characteristics and the risks that it carries; Sect. 3 describes methodological proposals to its mitigation formed on DEMO base; Sect. 4 describes the results, formulates questions for future research and concludes the paper (Table 1).

2   COSO Framework and Its Characteristics

Exceptions in the standards create situations in which the principles in the standards do not apply. Under a principles-based approach, it might not be possible to eliminate all exceptions [7]. However, the standard-setters believe that an objective of that approach should be to eliminate exceptions that are intended to achieve desired business results, which may obscure the underlying economics of the related transactions and events. To achieve that objective, it is necessary to resist pressures to provide exceptions in the standards. The stakeholders would need to accept the consequences of applying standards with fewer exceptions to the principles.

Under a principles-based approach, the key objectives of reporting are defined in the subject area and then the guidance explaining the objective is provided. While rules are sometimes unavoidable and the guidance should be sufficient to enable proper implementation of the principles, the intent is not to try to provide specific guidance or rules for every possible situation. A principles-based approach, while desirable, would require participants to exercise good professional judgment and resist the urge to seek specific answers and rulings on every implementation issue.

Management must evaluate and report on the effectiveness of internal control. Utilizing the COSO framework in internal auditing adds depth to the audit programs and constitutes the base for creating of broad internal audit work programs in compliance matter.

In addition to the principles, the 2013 Framework introduces 81 points of focus which represent typically important characteristics of principles that can be used to facilitate designing, implementing and conducting internal control. These are items, management can consider to determine if the principles are present and functioning. According to this Framework management is not required to separately evaluate whether each of the points of focus is in place to determine if the principles are present and functioning.

As COSO states, control activities are the actions established through policies and procedures that help ensure that management's directives to mitigate risks to the achievement of objectives are carried out. Control activities are performed at all levels of the entity, at various stages within business processes, and over the technology environment [4]. So, the process is a tool which supports a policy realization and the achieving of enterprise objectives.

Table 1. COSO internal control components and principles [4]

Component	Principle
Control environment	Principle 1: The organization demonstrates a commitment to integrity and ethical values. Principle 2: The board of directors demonstrates independence of management and exercises oversight for the development and performance of internal control. Principle 3: Management establishes, with board oversight, structures, reporting lines and appropriate authorities and responsibilities in the pursuit of objectives. Principle 4: The organization demonstrates a commitment to attract, develop, and retain competent individuals in alignment with objectives. Principle 5: The organization holds individuals accountable for their internal control responsibilities in the pursuit of objectives.
Risk assessment	Principle 6: The organization specifies objectives with sufficient clarity to enable the identification and assessment of risks relating to objectives. Principle 7: The organization identifies risks to the achievement of its objectives across the entity and analyzes risks as a basis for determining how the risks should be managed. Principle 8: The organization considers the potential of fraud in assessing risks to the achievement of objectives. Principle 9: The organization identifies and assesses changes that could significantly impact the system of internal control.
Control activities	Principle 10: The organization selects and develops control activities that contribute to the mitigation of risks to the achievement of objectives to acceptable levels. Principle 11: The organization selects and develops general control activities over technology to support the achievement of objectives. Principle 12: The organization deploys control activities as manifested in policies that establish what is expected and in relevant procedures to effect the policies.
Information and communication	Principle 13: The organization obtains or generates and uses relevant, quality information to support the functioning of other components of internal control. Principle 14: The organization internally communicates information, including objectives and responsibilities for internal control, necessary to support the functioning of other components of internal control. Principle 15: The organization communicates with external parties regarding matters affecting the functioning of other components of internal control.
Monitoring activities	Principle 16: The organization selects, develops, and performs ongoing and/or separate evaluations to ascertain whether the components of internal control are present and functioning. Principle 17: The organization evaluates and communicates internal control deficiencies in a timely manner to those parties responsible for taking corrective action, including senior management and the board of directors, as appropriate.

It was found [8] that, in the absence of time pressure, structured audit programs improve audit effectiveness, efficiency, and consistency. Disadvantages of a structured audit program include inflexibility (a fixed set of variables), mechanistic thinking (over-reliance on the structured audit program), audit inefficiency (not using sufficient available resources to successfully complete the task), and audit ineffectiveness (e.g., not detecting fraud when fraud is present).

3 Methodological DEMO-Based Proposal for COSO Improvement

The COSO Framework recognizes that while internal control provides reasonable assurance of achieving the entity's objectives, limitations do exist. Internal control cannot prevent bad judgment or decisions, or external events that can cause an organization to fail to achieve its operational goals.

Because many companies have implemented structured programs and the vast majority of companies, namely 82%, apply the COSO Framework, its revisions are likely to have widespread effects on Internal Audit compliance programs [7]. Whether required audit program adjustments will be substantial could depend on the flexibility of the program. Some organizations will view the new explicit principles and associated attributes as constituting additional requirements for assuring the compliance to COSO principles as additional requirements for its implementation and its subsequent assurance. The Institute of Internal Auditors suggests the 17 principles and 81 attributes will be perceived as a "checklist;" if the principles and attributes are employed in this manner, non-value added activities may increase regulation norms compliance burden [6].

Even being a very diffused framework its main limitation is that COSO document confuses and mixes up the framework (the organizational structures, policies, and arrangements put in place to promote, integrate and improve the management of risk) with the process used for risk management, particularly that used for risk assessment, risk treatment and monitor and review. They need to be thought of separately where the framework operates at an organizational level while the process is that which the framework seeks to integrate into all critical organizations processes where decisions are made.

We'll study the possibilities to eliminate these problems and defects from the enterprise engineering point of view and here the DEMO methodology [5] represents a valid support. The DEMO gives the analyst an understanding of the business processes of the organization, as well as the agents involved. Analysis of models built on DEMO allows the company to obtain detailed understanding of the processes of governance and cooperation and serves as a basis for business reengineering and information infrastructure development consistent with business requirements.

So, as one can see, the internal audit process in the company presupposes the existence of the following main actors:

- Board of Directors or external control body which initiate the audit process;
- Management;
- Auditor (internal or external one).

For the purpose of DEMO application which must be free of implementation aspect we can define these roles as "Board", "Decision-maker" and "Controller". An important element to apply the DEMO methodology is the Transaction Result Table (Table 2).

Table 2. The Transaction Result Table of internal audit procedure

Transaction type	Result type
T01 Control procedure start	Control procedure has been started
T02 Control results state	Control results have been stated
T03 Control results implementation	Control results have been implemented

Below we present one element of the Construction model, the Actor–Transaction Diagram. It expresses the main initiators and executors (CA) of the transactions individuated in the Transaction Result Table (Table 2).

As one can see, Auditor ("Controller") is the initiator of T3 because on the base of the results of his/her control activity the Management ("Decision-maker") elaborates the corrections in the business policies or processes.

All elements listed above represent a base for ontological model creation using the following models.

- the Construction Model (CM) which specifies the identified transaction types and the associated actor roles, as well as the information links between the actor roles and the information banks;
- the Process Model (PM) which contains, for every transaction type in the CM, the specific transaction pattern of the transaction type;
- the Action Model (AM) which specifies the action rules that serve as guidelines for the actors in dealing with their agenda;
- the State Model (SM) specifies the object classes and fact types, the result types, and the ontological coexistence rules.

So, internal audit involves human action, which introduces the possibility of errors in processing or judgment. Internal control can also be overridden by collusion among employees or coercion by top management.

These limitations preclude the board and management from having absolute assurance of the achievement of the entity's objectives - that is, internal control provides reasonable but not absolute assurance. Notwithstanding these inherent limitations, management should be aware of them when selecting, developing, and deploying controls that minimize, to the extent practical, these limitations.

The 2013 Framework [4] acknowledges that there are limitations related to a system of internal control. For example, certain events or conditions are beyond an organization's control, and no system of internal control will always do what it was designed to do. Controls are performed by people and are subject to human error, uncertainties inherent in judgment, management override, and their circumvention due to collusion. An effective system of internal control recognizes their inherent limitations and addresses ways to minimize these risks by the design, implementation, and conduct of the system of internal control. However, an effective system will not eliminate these risks. An effective system of internal control (and an effective system of internal control over financial reporting) provides reasonable assurance, not absolute assurance, that the entity will achieve its defined operating, reporting, and compliance objectives.

Below (Table 3) the advantages of different DEMO models and their application to COSO limitations according to [8] correction are presented:

Table 3. The corrective measures of different DEMO models

Limitation	DEMO model application
Suitability of objectives established as a precondition to internal control and the existence of external events beyond the organization's control	The Interaction Model (IAM), which represents the most compact ontological model of an enterprise, shows the boundary of the organization, as well as the interface transactions with actor roles in the environment [5]. This makes the IAM preeminently suitable for strategic alignment and objectives individuation. The IAM clearly presents the interface units of collaboration, namely complete transactions.
Reality that human judgment in decision making can be faulty and subject to bias and, more generally, human factor	To correct this limitation the apparatus of ontological maps based on DEMO and described in [2, 9, 10] can be used
Ability of management to override internal control	The IAM shows the ontological units of competence, authorization and responsibility. This may deliver a new view to human resource professionals, who have always struggled with finding the right chunks for the identification and classification of organizational functions. A comparison of the IAM with the current assignment of organizational functions to actor roles may provide the first ideas for improving it.
Ability of management, other personnel, and/or third parties to circum-vent controls through collusion	The Process Model facilitates these decisions considerably because it clearly shows that these side paths are either full-fledged transactions (in which original facts are created) or not. In the latter case, the approval turns out to be only a matter of being informed about or a situation of an unclear assignment of authority. It is also very well suited to forward the discussion about the assignment of organizational functions to actor roles.

4   Results and Conclusions

In this paper we introduce the approach to the most diffused audit framework COSO limitations mitigation by means of DEMO methodology which later be used in further research. The ontological approach expressed by means of DEMO represents a conceptual model that only shows the essence of an enterprise or a business process and is coherent, comprehensive, consistent and concise [5]. These properties allow to reduce the design costs and can be applied to the modelling of audit activity in its operative phase as well.

Thus, analysis of DEMO models provides decision makers with particular means of organizational transformations. Such choice unavoidably deals with information systems management and from such positions the use of the DEMO for both enterprise structure modeling and individuation of the most suitable information system use is quite advantageous. DEMO is easily reproducible, and it can be applied regardless of the business segment of the enterprise.

So, the future direction of the research could be constituted by the further extension of DEMO models application to COSO limitations differing it by components or principles, detailing the contents of different DEMO models according to COSO's components or principles. It may also constitute, besides some practical case analysis, the enlargement of DEMO application also to other Internal Audit frameworks generalizing in DEMO language all terminological notions proper to them. We also plan to extend our starting research point (a formal description of internal audit process) to the usage of DEMO means for the audit principles accomplishment verification.

The reported study was funded by Russian Fund of Basic Research according to the research project № 16-06-00300-a.

References

1. Arjoon, S.: Striking a balance between rules and principles-based approaches for effective governance: a risks-based approach. J. Bus. Ethics **68**(1), 53–82 (2006)
2. Babkin, E., Malyzhenkov, P.: Assessment of brand competences in a family business: a methodological proposal. In: Pergl, R., Molhanec, M., Babkin, E., Fosso Wamba, S. (eds.) EOMAS 2016. LNBIP, vol. 272, pp. 129–138. Springer, Cham (2016). doi:10.1007/978-3-319-49454-8_9
3. Challagalla, G., Murtha, B., Jaworski, B.: Marketing doctrine: a principles-based approach to guiding marketing decision making in firms. J. Market. **78**(4), 4–20 (2014)
4. COSO Internal Control — Integrated Framework Executive Summary (2013)
5. Dietz, J.L.G.: Enterprise Ontology: Theory and Methodology. Springer, Heidelberg (2006)
6. Institute of Internal Auditors, Responses to COSO's public exposure feedback questions (2012)
7. Kasey, M., Sanders, E., Scalan, G.: The potential impact of COSO internal control integrated framework revision on internal audit structured SOX work programs. Res. Account. Regul. **26**, 110–117 (2014)
8. Shaw, H.: The trouble with COSO. CFO Mag. **22**(4), 74–77 (2006)
9. Sergeev, A., Babkin, E.: Towards a formal approach to solution of ontological competence distribution problem. In: Pergl, R., Molhanec, M., Babkin, E., Fosso Wamba, S. (eds.) EOMAS 2016. LNBIP, vol. 272, pp. 84–97. Springer, Cham (2016). doi:10.1007/978-3-319-49454-8_6
10. Sergeev, A., Babkin, E.: Towards competence-based enterprise restructuring using ontologies. In: Aveiro, D., Pergl, R., Valenta, M. (eds.) EEWC 2015. LNBIP, vol. 211, pp. 34–46. Springer, Cham (2015). doi:10.1007/978-3-319-19297-0_3

VISI Re*visi*ted

Niek J. Pluijmert[1,2]([⊠])

[1] Radboud University, Comeniuslaan 4, 6525 Nijmegen, HP, The Netherlands
[2] INQA Quality Consultants B.V.,
Herman van Swaneveltplein 19, 3443 Woerden, HZ, The Netherlands
pluijmert@inqa.nl

Abstract. In this paper we investigate the use in practice of the VISI standard. The goal of the VISI standard is to arrange the cooperation of the parties in construction projects. Application of the VISI standard is not without troubles. The VISI standard is based on the DEMO methodology. We used the Hevner Three Cycle View and Sein's ADR to understand *what* and we used interviewing and process mining to understand *how* the development and use of VISI has been. We conclude that an overall cycle over the three cycle view is necessary to see that the right process is followed in using scientific knowledge to design artifacts that solve practical problems.

Keywords: VISI · DEMO · ISO standard · Enterprise engineering · Enterprise ontology · Process management · Action research · Design science research · Action design research · Construction sector · Large infrastructure projects

1 Introduction

In the Dutch construction sector for large infrastructure projects cooperation between parties is becoming ever more important. With 'parties' is understood all companies and principal(s) that realize an infrastructure object in one project. The number of parties involved in one infrastructure project has risen and responsibilities have shifted. In order to improve the cooperation, in 1998 the VISI[1] project was started. This resulted in 2003 in the VISI standard and the VISI standard resulted in 2012 in the ISO 29481 standard[2]. In the Netherlands the use of the VISI standard is widespread and since 2012 its use is mandatory. The application of the VISI standard is not without trouble, so the owner of the standard, CROW[3], wants a new version of the VISI standard that overcomes the problems.

---

[1] VISI is a registered trademark of CROW. It is an acronym of *Voorwaarden scheppen voor het invoeren van standaardisatie ICT in de GWW-sector*, which can best be translated to "Creating conditions for introducing standardization ICT in the infrastructure sector".

[2] ISO 29481-2 was prepared by Technical Committee ISO/TC59, Buildings and civil engineering works, Subcommittee SC 13, Organization of information about construction works.

[3] CROW is not-for-profit knowledge partner for (decentral) government, contractors and consultancy firms.

© Springer International Publishing AG 2017
D. Aveiro et al. (Eds.): EEWC 2017, LNBIP 284, pp. 74–81, 2017.
DOI: 10.1007/978-3-319-57955-9_6

CROW wants insights from science to be taken into account. In this article we investigate the problems and try to define an approach for solving the problems.

VISI is founded on the DEMO (Design and Engineering Methodology for Organisations [1]) methodology. The core of the VISI project was convinced that the theory of DEMO was the right one to apply, because it is founded on communication theory. This leads to the following questions:

- What is the process of design and implementation of the VISI standard?
- How do projects that follow the VISI standard for structuring communication, implement the VISI standard?
- How has the improvement cycle of the designed artifact VISI been?

The remainder of this article is structured as follows. In Sect. 2 we write about research approaches for IT. In Sect. 3 we describe the case study of the VISI project and the use of VISI in construction projects. In this section the results of the VISI project are described and how VISI is used in practice. In Sect. 4 we discuss the results of the VISI case study and draw conclusions.

2   Research Approach

In [2] we wrote about the three cycle view of design science (DS) from Hevner [3], see Fig. 1. The left cycle, the relevance cycle, are the issues described and analyzed in Subsects. 3.1 and 3.2 and discussed in Sect. 4. The VISI standard itself is a designed artifact (middle cycle). The DEMO methodology and process mining theory are for our situation part of the knowledge base. We wrote in [2] also about action research (AR) as most appropriate method to study social phenomena and as a way to respond faster to environment's demands. Sein in [4] proposes to combine AR and DS into a new method Action Design Research (ADR) in order to combine theory with practice and thinking with doing, see Fig. 2. Hevner has primarily a cyclic view and Sein has a staged view. Keeping in mind nowadays practices like Agile, Scrum and Lean that focus on delivering

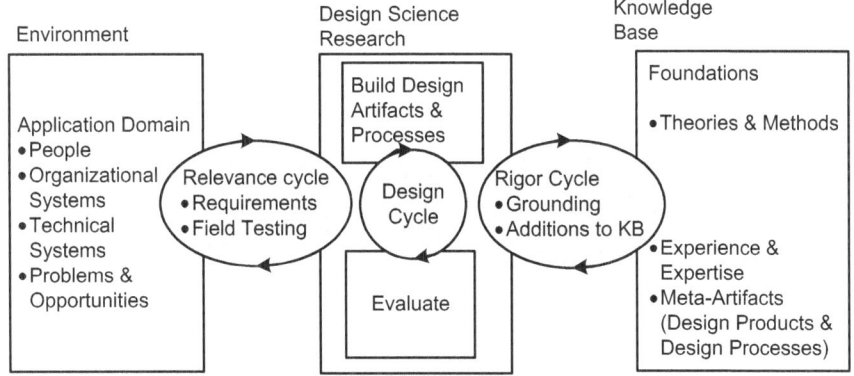

Fig. 1. Three cycle view of design science, Hevner [3]

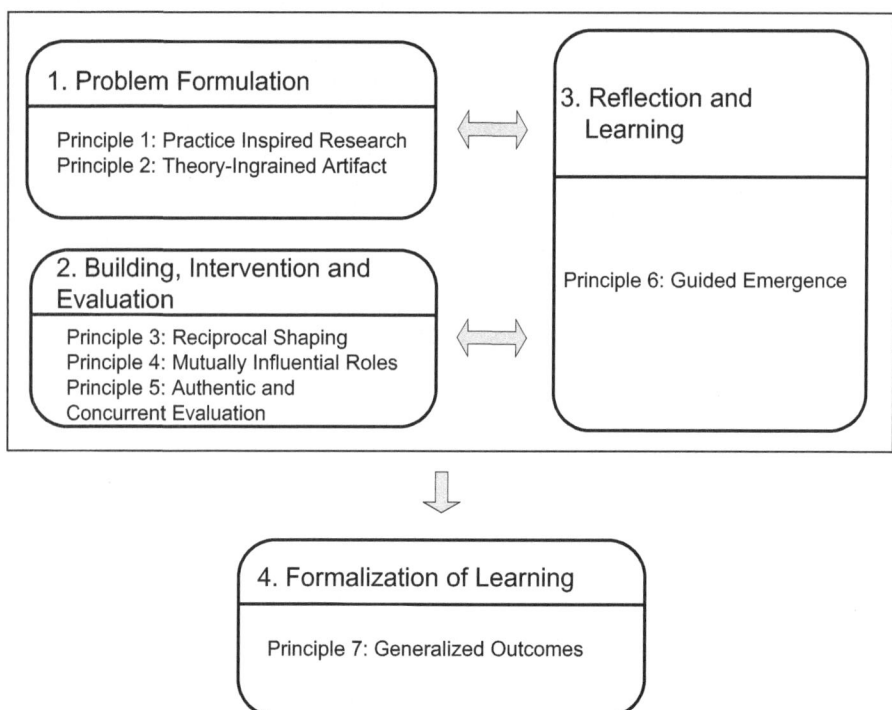

Fig. 2. Action design research (ADR) acc. Sein et al. [4]

added value fast, we will use Hevner and Sein to design our research method that is intended to deliver practical results fast and being thoroughly founded in science. The underlying epistemology we use is an interpretive one (see Myers [5], Orlikowski [6] and Chua [7]). In Action Research interviewing is the way of collecting data. We added *process mining* as a way to gather objectively data. Van der Aalst describes in [8] what process mining is. In our opinion the advantage of process mining with respect to VISI is that we have the facts about communication.

3   Case Study of VISI Development and VISI Use in Dutch Infrastructure Construction Sector

3.1   Introduction

In this section we describe and analyze in Subsect. 3.2 how the VISI standard was developed in the VISI project and in Subsect. 3.3 how it was used.

3.2   VISI Project

In the *investigation phase* (see Fig. 3 for a time line) the concept agreements of the VISI standard were developed. In this phase the choice for DEMO as

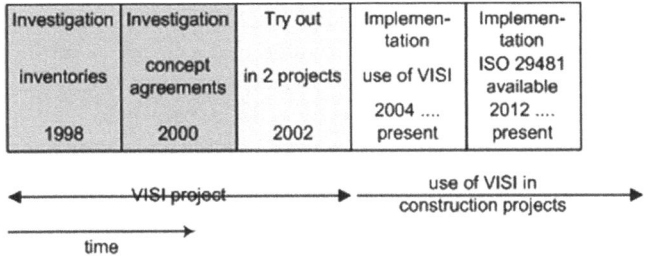

Fig. 3. Time line of the VISI project, extended with ISO standard; the VISI project ended in 2004, after that year the implementation started

underlying theory is made. The outcome consisted of the generic communication model (the existence of which was confirmed) and the table of contents of the VISI guideline.

In the *try-out phase* (2002–2004, see Fig. 3) the VISI model was verified in two infrastructure projects and the results were discussed in group decision sessions with the project managers of infrastructure projects. The project managers confirmed the results and that marks the coming into existence of the VISI standard: the first version of the VISI guideline.

VISI as standard can very well be compared with EDIFACT (see *e.g.* [9]). EDIFACT is a standard for electronic data interchange that provides a set of semantics and syntax rules to structure data, an interactive exchange protocol and standard messages which allow multi-country and multi-industry exchange. Just as EDIFACT, VISI is a business language so all stakeholders in the network are capable of understanding the responsibilities of the actors in executing business transactions. The VISI standard consists of:

- Interaction Chart or Framework for an infrastructure project. This chart is not formally part of the guideline and is a model of the communication in the construction sector. This model is independent of the way of cooperating that is contractually agreed.
- Principles and fundamentals. This consists of a description of parts of DEMO and of a way to design a framework. In this way of working, VISI deviates from the DEMO methodology while VISI recognizes other statuses and doesn't implement the complete transaction pattern.
- Specification of interaction framework and messages in XSD and XML formats.
- A software program, called Promoter, that generates a machine readable scheme in which all messages are defined, based on a framework. Software makers use this Promoter in developing VISI software.

3.3   Implementation of VISI

We looked at the developments in the standard and in the use of VISI. *Developments in the VISI-standard*

After the first version of VISI standard in 2003, there have been new releases in 2008, 2011, 2014, 2015, 2016. Governance and control is arranged: changes are prepared in the technical committee and approved in the steering group. In the governance and control organs science is not represented.

The developments in the standard have all been more (detailed) prescriptions for the form and content of the messages. The concept of *status* has been abandoned in 2014 release in favor of the concept of previous message determining the next message.

Table 1. Messages of process proposal for change

Proposal for change		
Acceptance of proposal	Acceptance of proposal, no financial consequences	Denial of proposal
Offer for change	Message of accomplishment	Withdrawal of proposal
Acceptance of offer for change	Acceptance of accomplishment	
Work completion statement		
Acceptance of work completion statement		

Developments in the use of VISI

For this purpose we analyzed the data of five projects that applied VISI software for communication. This data is analyzed with a process mining tool (ICRIS process miner) and for one case we interviewed the contract manager of the principal and the manufacturing engineer of the contractor. We used process mining mostly to learn about the processes of transactions, this is the messages that are sent consecutively from start (this is a message of a transaction before which is no other message) to finish (this is a message after which no other message follows). Table 1 gives the messages of an example process. From the names of the messages it is derived that it is a process of one transaction. In the simple example of Table 1 the DEMO pattern of request-promise-state-accept cannot be recognized. We see also a change of result in the transaction: first it is a proposal for a change, next it is an offer (proposal with a price) for a change and at last it is a work completion statement. And moreover a distinction is made between a proposal with and without financial consequences. With the process miner it is also possible to make a list of all start messages. From this list we learn that the use is for contractual changes, delivery of contractually agreed documents, work completion statements and the report of the constructors meeting. For all transactions it holds that they are between principal and contractor. The framework that is applied, is not documented, we can only reconstruct it with the process miner. We use here a simple example, in the data of other projects we saw much more complex processes that exist of up to ten transactions and those processes had the same properties as described here. New transaction types were about external judgment of a proposal or document. Judgment transactions are solely found within the organization of principal. From the interviews with contract manager and manufacturing engineer we learn that both were satisfied with how their communication was supported by VISI software. They delivered their wishes for the set up of the communication and between the two of them they had an appointment how to use it.

In the above paragraph we elaborated on some aspects of the use of VISI. Because of considerations for the length of this article, we summarize and do not go into detail here all our findings from interviews and process mining:

1. The use of the DEMO methodology is essential.
2. VISI applied the DEMO theory in a different way by defining other statuses than DEMO and during the use of VISI software the concept of status was abandoned.
3. VISI applied DEMO not completely.
4. VISI doesn't recognize the possibility of revoking a communication act.
5. VISI focuses on *coordination* solely, while coordination and production shouldn't be considered separately.
6. Several issues in the project context were important: the core group with its stable composition, the members of the core having decision power, commitment of top management, use of a participative approach towards project managers of construction projects.
7. After the first release of VISI standard the shift towards an IT based approach for defining and supporting the communication scheme of a project (project specific framework).
8. Users of VISI-software in a construction project are satisfied how it supports the communication.

4   Discussion and Conclusions

In this section we discuss the findings from the perspective of DS and ADR.

First from the DS-perspective. Item 1 confirms that the rigor cycle has been walked through correctly. Items 2, 3 and 5 state that in the design cycle it is decided to deviate from the theory because the project members decided that this was the best they could do to get the standard accepted and applied in the construction sector. So, here is a decision taken that should have been tested in a relevance cycle. Item 4 is an issue that was still in development in DEMO, so this could have been an addition to the knowledge base. Item 6 has not much to do with the three cycles of design science but is an important condition for a successful (design) project! Items 7 and 8 are facts from the relevance cycle. The application of the VISI standard in construction projects is made with an IT-perspective and mostly by IT-people.

From the ADR-perspective (see Fig. 2) we see that the investigation phase was about Problem Formulation. At the end of investigation phase and during try-out phase Building, Intervention and Evaluation is recognized because in those phases VISI standard was defined and built. Also Reflection and Learning is recognized, because inventories were made and also the solution was found appropriate for application in other sectors than building. In our data we didn't find so much that points to Formalization of Learning.

In the implementation phase, when VISI is used in construction projects for communication between principal and contractor, design cycles and ADR-stages are not so well recognizable. There is a cycle of 1.5 year from determination of the content of a new release till the availability of the adapted software for use. In terms of design science, it seems that the relevance and design cycle are walked through regularly but only to the extent that software has been developed. In terms of ADR, it is only part of stage 2, Building, Intervention an Evaluation. It is good te repeat that VISI standard is about 2 things:

1. a theory about communication between people (Principals and fundamentals).
2. a specification for software that supports communication between people with digital messages (Specification of interaction framework and messages).

It seems that during the use of VISI (see Fig. 3) the development of VISI standard has not been based on research and science anymore. But referring in Sect. 1 to encountered problems, research and science are necessary and there should be a cycle that takes into consideration whether all aspects are dealt with properly, that the right process is followed and that the organization can deal with the implied change. This last aspect is different from the distinction in IT-dominant and organization-dominant BIE that Sein makes in [4, p. 42]. In Sein [4] it is about the content of the artifact to be designed, we aim at the impact of a change that an organization can handle or the answer to the question whether the organization will accept a certain change. We define this as a cycle that implies environment, design science research and knowledge base. By constantly taking into account where (in which cycle) what has to be done, we come ever closer to the desired result. In [11] Argyris *et al.* describe this aspect. Argyris *et al.* call this double loop learning. Single loop learning is design the artifact and improve it, while double loop learning also takes the followed process as subject for improvement. In Fig. 4 we represent this by a spiral over the three cycles according to Hefner. In ADR (Fig. 2) this aspect can be imagined implicitly in the double arrows, but it would be more clear if it was represented by a separate rounded rectangle called management of change. Double loop learning takes both the *what* (three cycles) and the *how* (organizing and controlling the process or project management) into account. In [2] we wrote about the participative project approach of Mulder [10] that has a large added value in AR research because of the coherence between this approach and decision making in an organization. Such an approach could help prevent the problems that are encountered in the application of VISI.

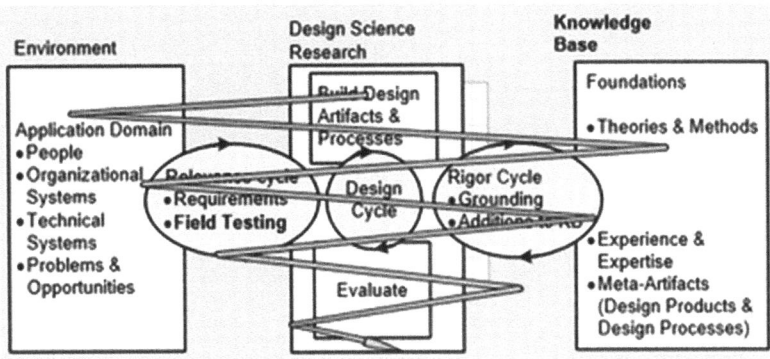

Fig. 4. Design science cycles with action research spiral

References

1. Dietz, J.L.: Enterprise Ontology Theory and Methodology. Springer, Berlin (2006)
2. Pluijmert, N.J., Molnar, W.A., Proper, H.A.: Research approach in enterprise engineering: a matter of engineering. In: Franch, X., Soffer, P. (eds.) CAiSE 2013. LNBIP, vol. 148, pp. 73–86. Springer, Heidelberg (2013). doi:10.1007/978-3-642-38490-5_6
3. Hevner, A.R.: A three cycle view of design science research. Scand. J. Inf. Syst. **19**(2), 87–92 (2007)
4. Sein, M.K., Henfridsson, O., Purao, S., Rossi, M., Lindgren, R.: Action design research. MIS Q. **35**(1), 37–56 (2011)
5. Myers, M.D.: Qualitative research in information systems. MIS Q. **21**(2), 241–243 (1997)
6. Orlikowski, W.J., Baroudi, J.J.: Studying information technology in organizations: research approaches and assumptions. Inf. Syst. Res. **2**(1), 1–28 (1991)
7. Chua, W.F.: Radical developments in accounting thought. Account. Rev. **61**(4), 601–632 (1986)
8. Van der Aalst, W.: Process Mining Discovery, Conformance and Enhancement of Business Processes. Springer, Heidelberg (2011)
9. https://en.wikipedia.org/wiki/EDIFACT: United Nations/Electronic Data Interchange for Administration, Commerce and Transport (UN/EDIFACT) is the international EDI standard developed under the United Nations
10. Mulder, J.B.F.: Rapid enterprise design. Ph.D. thesis, Technical University Delft (2006)
11. Argyris, C., Putnam, R., McLain Smith, D.: Action Science. Concepts, Methods, Skills for Research and Intervention. Jossey-Bass Publishers, San Francisco (1985)

Business Processes

Converting DEMO PSI Transaction Pattern into BPMN: A Complete Method

Ondřej Mráz, Pavel Náplava, Robert Pergl$^{(\boxtimes)}$, and Marek Skotnica

Czech Technical University in Prague, Prague, Czech Republic
mraz.ondra@gmail.com, naplava@fel.cvut.cz,
{perglr,skotnicam}@fit.cvut.cz

Abstract. The goal of this paper is to contribute to efforts of improving the Business Process Modelling (BPM) practice. We present an original method for converting 0enterprise ontology Design & Engineering Method for Organisations (DEMO) process models into a BPMN 2.0 notation. By this approach, we are able to mitigate certain methodological deficiencies of BPMN. The method exhibits the following qualities: Implementation of the complete transaction pattern formulated by the PSI-theory, correct managing of multiple child transaction instances, and executability of the resulting BPMN model.

Keywords: PSI-theory · BPMN · DEMO · Business Process Modelling · Enterprise ontology · Conceptual modelling

1 Introduction

BPMN (Business Process Model and Notation) [1] is a graphical notation that is used for modelling business processes. Key characteristics of BPMN are simplicity of the underlying theory (flowchart), standardised notation and a large number of tools. This makes BPMN one of the most wide-spread process modelling notation in practice, in spite of its limitations and flaws. BPMN offers three different types of diagrams: Choreography, Conversation and Collaboration diagrams. For this work, only the Collaboration Diagram will be considered. This diagram expresses the process flow in achieving participants' goals.

One of the BPMN weaknesses is the absence of a methodology for constructing diagrams, which is addressed for example by Silver [2]. Nevertheless, the design freedom is still too broad, which results in different modelling styles of individual analysts and different models depicting the same situation, which complicates enterprise engineering tasks like mergers and reorganisations.

DEMO (Design & Engineering Method for Organisations) [3] is a leading modelling method used in the discipline of Enterprise Engineering [4] based on deep and sound theoretical basis (the PSI-theory) and high ontological relevance. Its benefits for the practical use has been proven, as documented for example in [5] or [6]. It does not limit itself just to process modelling, but it also deals

© Springer International Publishing AG 2017
D. Aveiro et al. (Eds.): EEWC 2017, LNBIP 284, pp. 85–98, 2017.
DOI: 10.1007/978-3-319-57955-9_7

with capturing structural (factual) knowledge and business rules, thus delivering a complete enterprise ontology exhibiting certain criteria (C4E). However, compared to BPMN, DEMO is still a niche approach and relatively demanding to master. Also, a limited number of tools is available today.

For a brief description of DEMO, we take a help of Op't Land and Dietz [5]:

A complete, so-called essential model of an organization consists of four aspect models: *Construction Model (CM)*, *Process Model (PM)*, *Action Model (AM)*, and *State Model (SM)*. The CM specifies the composition, the environment and the structure of the organization. It contains the identified *transaction types*, the associated *actor roles* as well as the information links between actor roles and transaction banks (the conceptual containers of the process history). The PM details each transaction type according to the *universal transaction pattern*. In addition, it shows the structure of the identified business processes, which are trees of transactions. The AM specifies the imperatively formulated *business rules* that serve as guidelines for the actors in dealing with their agenda. The SM specifies the *object classes*, the *fact types* and the declarative formulations of the *business rules*.

The DEMO Process Model reveals details of the transactions with the respect to universal transaction pattern. The basis is the "happy flow" consisting of `request`, `promise`, `state` and `accept`, which is also called the *basic transaction pattern*. In the so-called *standard transaction pattern* (not depicted), `decline` may happen instead of `promise` and `reject` may happen instead of accept. Then, a new attempt may be made, or `quit`, resp. `stop` may end the transaction unsuccessfully. Real situations may become even more complicated, which is addressed by the *complete transaction pattern* in Fig. 1. It incorporates the notion of *revocation* – an actor may want to "take back" their act done before[1]. If that is allowed by the other party, the transaction "rolls back" to the desired state.

The logic of the complete transaction pattern is automatically included in all DEMO transactions, which is one of the reasons why the models are compact.

The main goal of this paper is to combine the simplicity of the BPMN and ontological qualities of the DEMO. The result is the method that converts enterprise ontology Design & Engineering Method for Organisations (DEMO) process models into a BPMN 2.0 notation. This approach mitigates the mentioned absence of a sound methodological approach for BPMN. The BPMN models resulting from the described method converge, similarly to DEMO, to one essential model, thus eliminating different modelling styles of individual analysts leading to comparable models. Our other requirements are: implementation of the complete transaction pattern formulated by the PSI-theory, correct managing of multiple child transaction instances, and executability of the resulting BPMN model.

We start the paper by the discussion of the related work of efforts of improving BPM and BPMN, specifically the approaches based on applying the enterprise-engineering rigour (Sect. 2). We then briefly present results of a comparative analysis of DEMO and BPMN (Sect. 3), which led to formulating our

---

[1] In the DEMO theory, nothing can disappear, so the original fact remains in the fact bank, however, the transaction flow is changed.

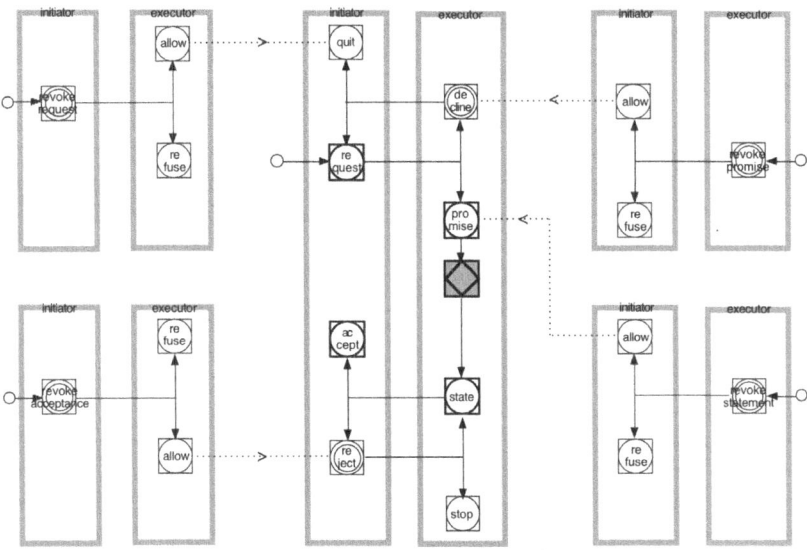

Fig. 1. DEMO complete transaction pattern [7]

method of conversion (Sect. 4). We demonstrate the method on an example (Sect. 5). Finally, we discuss the result and formulate conclusions (Sect. 6).

2   Related Work - Improving BPM and BPMN

A poor ontological quality of BPMN is generally known and documented [8]. The most practised remedy is exercising a methodological approach like the one proposed by Silver [2], who distinguishes three levels of BPMN: (i) Descriptive, (ii) Analytical and (iii) Executable and proposes several analysis patterns and anti-patterns.

The discipline of enterprise engineering (EE) [4] brought about a rigorous approach of building an enterprise ontology (EO) [3], DEMO being its modelling method. There are several foundational EE theories, the most notable being the PSI-theory. As on of the central concerns of EE is the business process management, the effort to apply EE theories (EET) to the existing (less formal approaches) is promising. The efforts in this area are twofold:

1. Applying EET for analysis of existing BPMN models of business processes: for example [9–11].
2. Enhancing the formal foundations of BPMN by EET: for example [9,10,12,13]

2.1   Applying EET for Analysis of Existing BPMN Models of Business Processes

Caetano et al. showed that applying the DEMO PSI-theory to improve business process modelling deserves attention [9]. The authors started by analysing existing BPMN models and identified missing DEMO transaction pattern steps in these models. It had been determined or each BPMN activity from the analysed models, whether this activity is an ontological, infological or data-logical part of a transaction. It had been also determined, which part of the transaction pattern each activity represents. Next, the authors created an ATD and a PSD diagram of DEMO and using a PSD diagram, they enriched existing BPMN models by adding missing parts of the transaction pattern into the BPMN models.

In the second part, the authors present results of applying this method to analysis of existing BPMN models of key processes of a big organization (more than 500 activities and 60 actors). The authors identified numerous missing act types in the original BPMN models. The results from this analysis were: (i) 25% of production C-acts missing in the original BPMN model, (ii) 25% of request C-acts missing in the original BPMN model, (iii) 50% of promise C-acts missing in the original BPMN model, (iv) 25% of state C-acts missing in the original BPMN model, (v) 40% of accept C-acts missing in the original BPMN model.

Results reported by Pergl and Náplava for an academic institution [11] state reduction of DEMO essential models complexity to 21% of the original BPMN size and several model quality improvements similar to [9].

2.2   Enhancing the Formal Foundations of BPMN by EET

These efforts aim to precisely express the EE ontological constructs using the standard BPMN notation. Two approaches have been followed. The first is to enhance the BPMN models by adding the missing C-(F)acts and other constructs from the PSI-theory. Caetano [9] is an example of this method.

The second way is generating BPMN models from the DEMO models. This method was discussed in the diploma theses [13], from which the approach in this paper was designed.

3   Analysis of DEMO and BPMN

Here follows observations of comparing various aspects of DEMO with respect to BPMN, from which follows the conversion principles and decisions made. These were formulated based on the DEMO theory axioms and models definitions related to the BPMN elements definitions, as introduced in Sect. 1.

- Similar parts of methods that can be simply transformed from the DEMO to BPMN:
  - The *Process Structure Diagram (PSD)* of DEMO contains process information, which can be related to a BPMN process diagram.
  - The *Action Model (AM)* of DEMO expresses complex decision rules for Coordination acts (C-acts)[2]. The contained information can be used for branching in BPMN.

---

[2] Apart from containing all the information from the other models.

- BPMN does not distinguish the three key human abilities (forma, informa, performa), however applying this distinction can be introduced straightforwardly, as shown for example in [11]. As this concern is orthogonal to our effort, we do not discuss the distinction axiom here.
- Related to the point above, the (atomic) actor roles in DEMO are executors of exactly one transaction, while swimlanes may contain many different actions.
- Different parts of methods that require deep analysis before transformation from DEMO to BPMN:
  - The DEMO *Transaction Axiom* concept does not exist in BPMN. Only happy flows and the most obvious unhappy flows are expressed in models.
  - The *Object Fact Diagram (OFD)* being a factual model does not have an analogy in BPMN.
  - DEMO and BPMN employ different execution models. While BPMN is flow-based, DEMO operates on the basis of a so-called CRISP model [3], which may be characterised as an event-driven, or more precisely, an agenda-driven execution model.
  - The *Construction Model (CM)* of DEMO is an abstraction that does not specify process, it provides just structural information.

4   Converting DEMO into BPMN

The goal is to convert the complete transaction axiom into BPMN, including all revoke types. Sections 4.1 to 4.4 describe all the necessary pieces and Sect. 4.7 presents the result. We used BPMN 2.0 and leveraged the newly available *Data Store* construct.

4.1   C-acts

C-acts are essentially activities that take place in order specified by the transaction pattern. BPMN has the concept of *activities* and the order is specified by *sequence flows*. As C-acts are atomic, the appropriate activity type is *task*.

4.2   C-facts

As mentioned in Sect. 1, a C-fact becomes existent in the world as a consequence of performing a C-act. Heller in his thesis [13] lists three possibilities of expressing C-facts using BPMN:

1. Not explicitly expressed – the existence of the fact-C is not explicitly expressed. It is indirectly realised by a sequence flow. This option is sufficient if revokes are not considered (see further).
2. Using a BPMN message – the actor, who performs the given C-act sends a BPMN message with the C-fact to the other actor (transaction participant).

3. Using a BPMN signal – the actor, who performs the given C-act emits a BPMN signal on creating a C-fact. This has the benefit that apart from the other actor, any other actor may subscribe to the signal reception, which is aligned with the PSI-theory, where facts are present in the world, not only in the transaction, thus available also outside the transaction (modelled by interstriction links).

However, under a closer consideration, none of the above solutions are completely sufficient for a correct handling of revokes. For each revoke act, the PSI-theory specifies a certain *state* in which the transaction must be. The state is formulated like "X or further": request(ed) or further, promise(d) or further and so on. This is why we decided for another representation: the BPMN 2.0 *data store*, into which the state of the transaction is stored. This data store is connected to every C-act activity by an association.

4.3 P-(F)acts

It is not necessary to store information about them creating a P-(f)act into the data store, because they can be derived from C-(f)acts: According to the PSI-theory, the P-fact starts to exist based on acceptance of the product, so P-(f)acts can be expressed by an activity only. If need be (optimisation of an implementation), they can be stored similarly to the C-(f)acts described above.

4.4 Actors

Swimlanes in BPMN are isomorphic to actors in DEMO [10]. BPMN lacks a higher abstraction level of actor roles, being the logical sum of responsibility, authority and competence necessary to carry out the product [3]. There are generally two approaches: (i) abstracting the swimlanes to actor roles (like Decider or Concluder), (ii) remaining on the BPMN's low level of abstraction and using swimlanes to represent actors – company functional roles – like CEO or specific people like Jane.

Another possibility for representing actor roles is using BPMN *pools*, where each pool represents one actor. The resulting BPMN models will be very similar to models using swimlanes, however we have not chosen this representation because: (i) The correspondence of actor roles and transactions is not explicit, (ii) sequence flow cannot be used between pools, which would result in using messages, further complicating the diagrams.

4.5 The Composition Axiom

A composition of transactions may be dealt with in two ways: (i) to model all the transactions in one diagram, (ii) to separate diagram for every transaction. Generally, both approaches are valid, but (ii) may lead to huge diagrams, as can be seen in Figs. 8 and 9. As (ii) guarantees the limit of the diagram size, we preferred it. On the other hand, it may render understanding of the big picture harder.

We propose the following 2-part expression of the composite axiom:

1. **Launching a child transaction** in a specific place in the parent transaction. The child transaction must be started just after creating a specific C-fact. A *message-throwing event* may be used in case of initiating a single child transaction. In case of firing multiple child transactions, signals are appropriate, similar to the C-acts above. Moreover, it is needed to ensure the multiplicity. In case that it is greater than one, we need to initiate several child transaction instances. This is achieved either by using a *cycle* for creating child transactions or a *loop activity*. Modelling by cycle (Fig. 2) means, that the model contains an activity counting, how many times the activity was run. After this activity, there is a gateway. If the counter has not reached the number of child transaction instances to spawn, the process goes into message throwing event to start a child transaction instance and then the process returns to the counting activity. This happens $0...N$ times, as required. When multiplicity is modelled by a loop activity (Fig. 3), the activity is in the form of a subprocess (with parallel loop), which sends a signal[3] that starts a child transaction. In the examples described below, the first (counter) variant is used because the model is more explicit. At the same time for models with a multitude of child transactions, the more concise loop variant is recommended. Also, from the execution point, the implementation variant may be driven by the vendor, as correlation of instances must be ensured (more discussed in Sect. 4.8).

2. **Blocking execution of the parent process** until the child process has not reached the given state (creating a C-fact being waited on). This blocking can be realized by a BPMN *catching event condition* in the parent process waiting for a specific condition before the given C-act. Here, a conditional event must be used instead of a signal event, as we do not wait just for a signal, but for a specific instance in case of multiple child transaction instances. This situation is modelled in Fig. 4. Again, specific vendor correlation techniques may apply (Sect. 4.8).

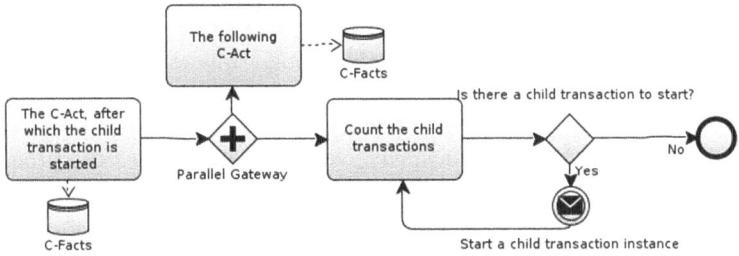

Fig. 2. Launching child transactions by using counter

---

[3] We cannot use a message send in this situation, because the encapsulation would be violated.

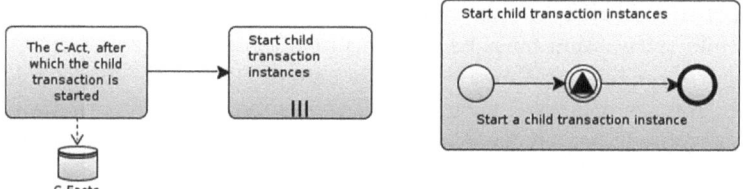

Fig. 3. Launching child transactions by using loop

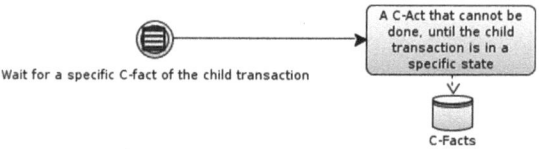

Fig. 4. Waiting for a child transaction

4.6 Revokes

Revokes are the most challenging part of the conversion. Let us present the challenges and how we dealt with them:

- A revoke must be applied on a specific instance of the transaction; in a certain time, there can be several parallel transaction instances running. This must be ensured by the BPMN system (Sect. 4.8).
- A revoke can be fired independently on the running main process. Can be modelled straightforward, as BPMN allows several independent start events.
- A revoke can be fired only if the transaction is in an allowed state. This we ensure by an activity checking the state of the transaction, which was previously stored into a data store.
- When revoking a C-fact, after which a child transaction has been started, the child transaction must be completely revoked. This is done by calling a compensation throwing event by the revoke, followed by performing the compensation activity by the corresponding parent transaction.
- In the process flow, there can happen a situation, that a P-fact was already created (the P-act has been finished), while a revoke moves the process to a state preceding performing the P-act. In this case, it is necessary to "throw-away" the P-fact. We solve this using a BPMN compensation element and the respective compensation activity, similarly to the previous point.
- A revoke must be initiated by the actor who performed the respective C-act to be revoked. This is ensured by using the same identifier for the swimlane of the actor role initiating the revoke as for the actor role of the respective transaction.

A revoke works in the following steps according to the transaction pattern. First, the revoking actor asks the other actor for granting the revoke. The other

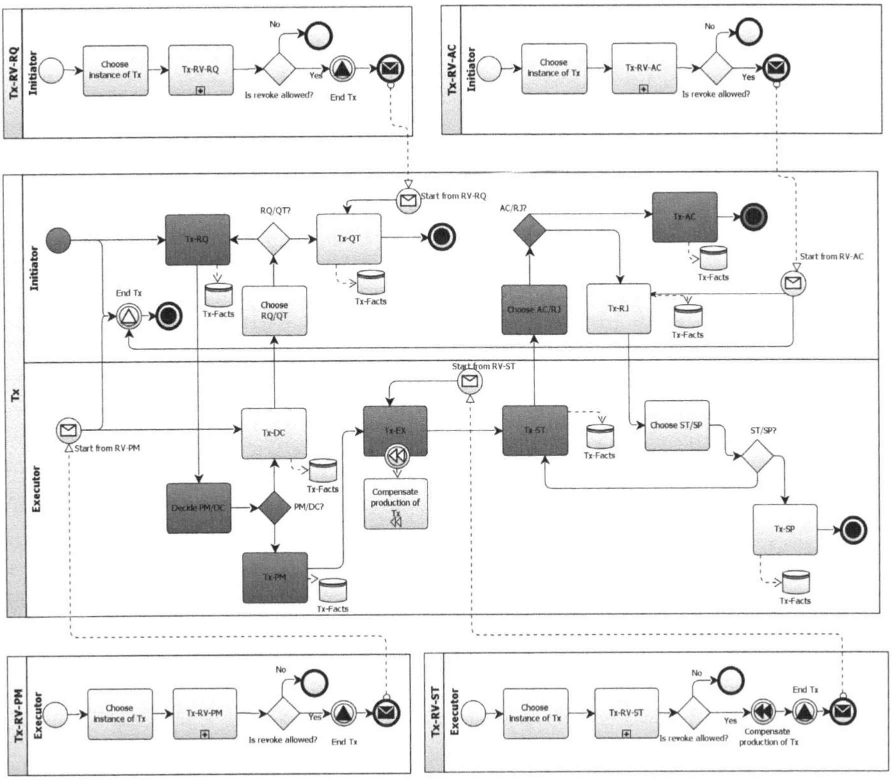

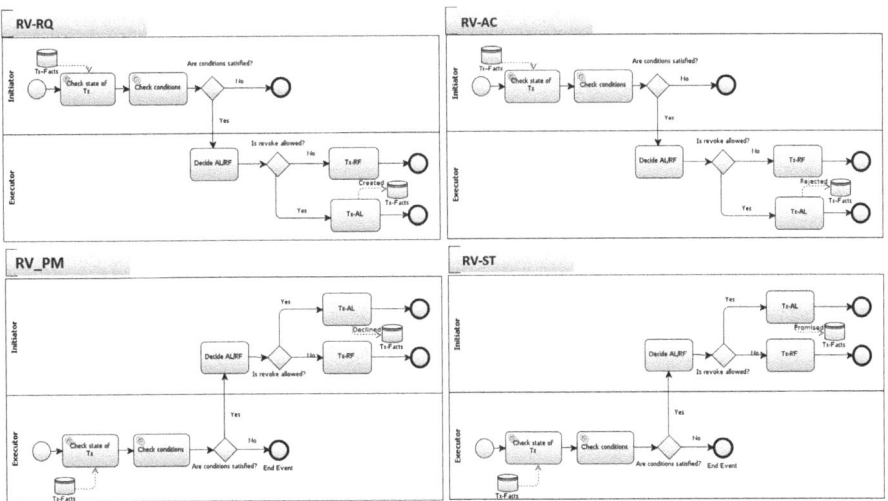

Fig. 5. Transaction in BPMN, Happy flow is marked by green colour (Color figure online)

Fig. 6. Revokes in BPMN

actor allows or refuses. If the revoke is allowed, the main process returns to the appropriate state. We model this by using simple BPMN *subprocess* with a set of appropriate activities (Fig. 6).

4.7   The Resulting BPMN Model

The complete transaction pattern described by the BPMN notation illustrates Fig. 5[4]. Although it describes only one transaction, it is very complex and complicated. As it is presented in Sect. 5 and discussed in Sect. 6, models containing more than one transaction are not easily readable by usual readers and it is recommended to use them for the process execution in BPM systems.

4.8   The Execution

Apart from documentation purposes, BPMN models can be simulated and/or executed. While designing the conversion, we tried to make the resulting BPMN model precisely following the required behaviour. Unfortunately, the BPMN standard does not specify the execution implementation details. Each company developing BPM system (system for modelling, simulation and execution of processes), as Intalio, BizAgi or IBM, has their specific implementation, which requires various additional modelling and programming steps necessary to make the model executable. At the same time, some of the BPMN constructs may not be supported or they are implemented differently. All these aspects must be taken into consideration for turning the resulting BPMN models into an executable form. Generally, here are the things that must be implemented:

– Agenda handling. The possibility to start a process and providing a "task inbox" of the required reactions on the originating C-facts. This requires developing some sort of user interface (UI).
– Allowing the participants to make their choices. Again, some sort of UI solves this. Also, some choices may be determined by complex facts evaluation specified in the Action Model. There are two possible approaches:
  1. Leaving the evaluation to users, which means the users have the rules in their head or consult the Action Model or any other codification of the rules.
  2. Programming the BPM system to (help) evaluate the rules. The extent to which the automation may happen depends on the possibilities of the BPM system used and also on the context (the availability of the necessary data in the company technological systems and their accessibility).
– Signals handling.
– Implementation of reading and writing data to data stores.
– Instance matching. Specific instances of transactions must be matched in some situations as child transactions (Sect. 4.5) and revokes (Sect. 4.6). Intalio and Oracle call this concept a "correlation".

---

[4] This and the following models may not be legible in the printed version. We recommend obtaining the electronic (zoomable) version. The source models may be downloaded from https://ccmi.fit.cvut.cz/methodologies/bpmn/.

5   Example – Case Voley

As an example for the demonstration of our method, the traditional Case Voley example [7] was selected because of its simplicity, yet including the substantial constructs. In Fig. 7 there is the OCD diagram of this example.

The process has two transactions and three actors. The transformed BPMN model converted by the described method is in Figs. 8 and 9. Subprocesses depicted in Fig. 6 are not shown here, as they are generally the same.

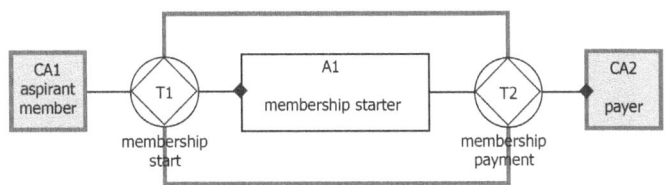

Fig. 7. OCD of Case Voley [7]

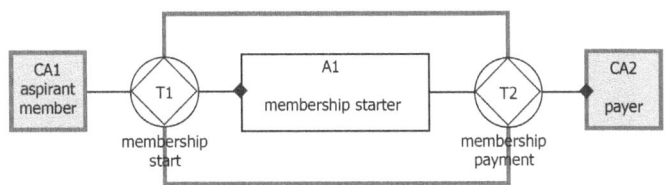

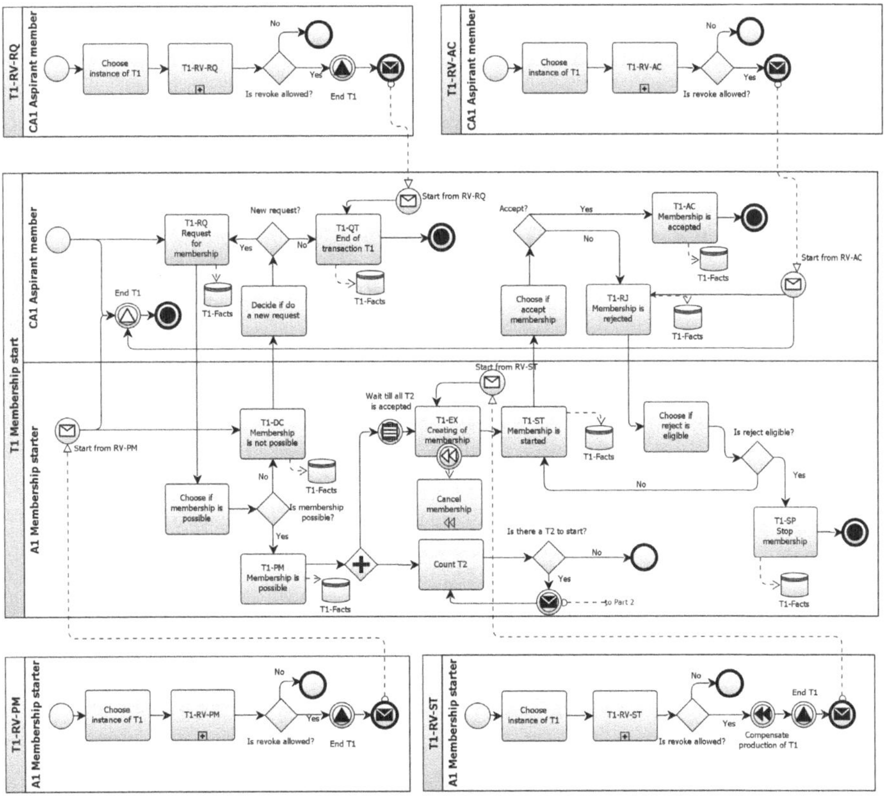

Fig. 8. Case Voley converted into BPMN – part 1

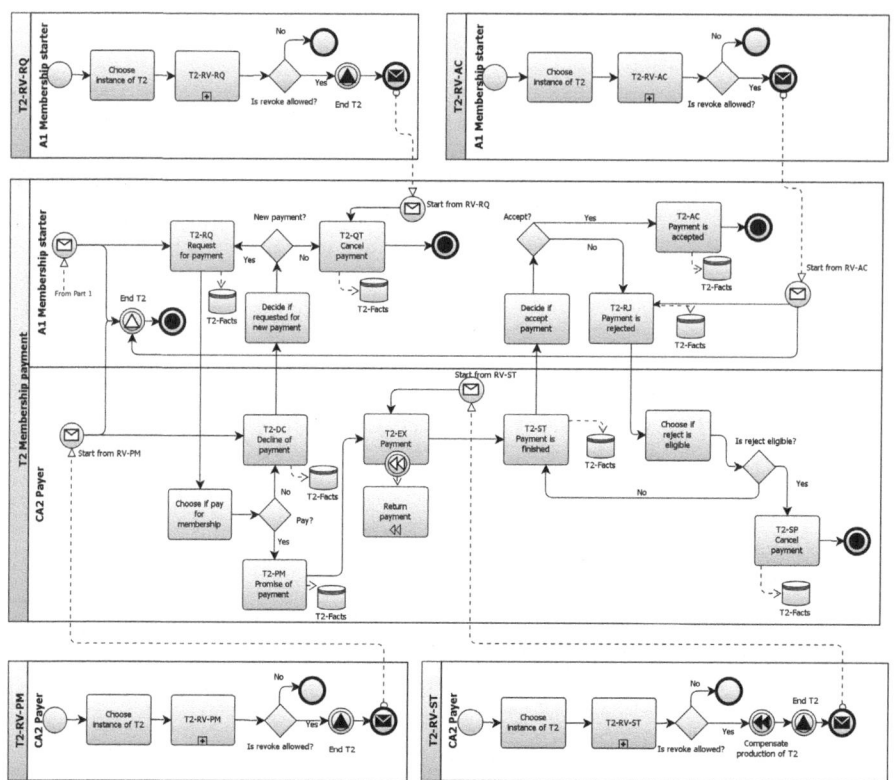

Fig. 9. Case Voley converted into BPMN – part 2

6   Discussion and Conclusions

The limitation of typical BPMN models from the view of the PSI-theory lie in their limited expression of reactions to unexpected situations. Many situations like decline, reject and especially revokes are not covered in the models, which causes operation troubles. The presented conversion method offers a remedy to this by bringing the *complete transaction pattern into BPMN*, which means including all revokes. Moreover, compared to the previous efforts, our method deals with spawning of *multiple child transaction instances* (initiation links with multiplicity $\neq 1$) and waiting for them in the parent transaction (waiting links with multiplicity $\neq 1$). Also, the resulting models are *executable*.

As for the DEMO models covered, the described conversion method covers the Construction Model plus the Process Model. Based on a concrete BPM system implementation, decision rules contained in the Action Model can be incorporated in the respective activities, as described in Sect. 4.8, which is also true for rules from the State Model.

The concept of interstriction has not been discussed, however a keen reader has probably realised that whenever an actor in its activity needs a specific information from another transaction, it is simply modelled by accessing the respective transaction data store.

The example shows that in spite of the simplicity of the DEMO model involved, the resulting BPMN model is complex. The reason is mostly the complete transaction pattern, which covers all the possible situations according to the PSI-theory. The question arises about the human readability. There are several points to this topic:

1. In practice, the model may be made smaller by leaving out the parts, which are not applicable (which means they (almost) never happen). These are typically the revoke patterns.
2. Yet, for complex models the resulting size may remain still unmanageable. In this case it would be advisable to cut the model into smaller pieces using some sort of decomposition and/or *link* BPMN elements. The concrete way how to do this may be explored in a future research.
3. It is questionable whether a human readability is required. If one wants human-readable diagrams according to the PSI-theory, the DEMO diagrams are the solution, as they have been tailored to it. It may be the case that learning and applying them comes at a lower cost than forcing the diagrams into a BPMN notation, just because "BPMN is the standard".

Our stance is that the greatest possibilities of our method lie in *machine readability*, which means generating BPMN models that can be fed into a BPMN execution system to implement an automated workflow that is able to react to every possible situation specified by the complete transaction pattern, not just a typical BPMN "happy path with a bit of branching".

Apart from converting the DEMO models, the conversion may be applied also for analysis of existing BPMN models of business processes as described in Sect. 2.1. The way of working would be to transform the BPMN models into DEMO and then generate the "supercharged" BPMN version by converting them back using our method.

As for the future work, a verification on a bigger models from practice is necessary. As such conversion will not be feasible by hand, an implementation of the conversion automation will be required.

Acknowledgements. This research has been funded by CTU SGS grant No. SGS16/120/OHK3/1T/18. The authors wish to deeply thank ForMetis BV and especially Dr. Steven van Kervel for the kind support of this research.

References

1. OMG: OMG: Business Process Model and Notation (BPMN) Version 2.0
2. Silver, B.: BPMN Method and Style, 2nd edn. with BPMN Implementer's Guide: A Structured Approach for Business Process Modeling and Implementation Using BPMN 2.0. Cody-Cassidy Press, New York, October 2011
3. Dietz, J.L.G.: Enterprise Ontology: Theory and Methodology. Springer, Berlin (2006)

4. Dietz, J.L.G., Hoogervorst, J.A.P., Albani, A., Aveiro, D., Babkin, E., Barjis, J., Caetano, A., Huysmans, P., Iijima, J., Kervel, S.J.V.: The discipline of enterprise engineering. Int. J. Organ. Des. Eng. **3**(1), 86–114 (2013)

5. Op 't Land, M., Dietz, J.L.G.: Benefits of enterprise ontology in governing complex enterprise transformations. In: Albani, A., Aveiro, D., Barjis, J. (eds.) EEWC 2012. LNBIP, vol. 110, pp. 77–92. Springer, Heidelberg (2012). doi:10. 1007/978-3-642-29903-2_6

6. Décosse, C., Molnar, W.A., Proper, H.A.: What does DEMO do? A qualitative analysis about demo in practice: founders, modellers and beneficiaries. In: Aveiro, D., Tribolet, J., Gouveia, D. (eds.) EEWC 2014. LNBIP, vol. 174, pp. 16–30. Springer, Cham (2014). doi:10.1007/978-3-319-06505-2_2

7. Dietz, J.L.: The Essence of Organization - An Introduction to Enterprise Engineering. Sapio bv, Voorburg (2012)

8. Guizzardi, G., Wagner, G.: Can BPMN be used for making simulation models? In: Barjis, J., Eldabi, T., Gupta, A. (eds.) EOMAS 2011. LNBIP, vol. 88, pp. 100–115. Springer, Heidelberg (2011). doi:10.1007/978-3-642-24175-8_8

9. Caetano, A., Assis, A., Borbinha, J., Tribolet, J.: An application of the Ψ-theory to the analysis of business process models. In: Poels, G. (ed.) CONFENIS 2012. LNBIP, vol. 139, pp. 258–267. Springer, Heidelberg (2013). doi:10.1007/978-3-642-36611-6_24

10. Nuffel, D., Mulder, H., Kervel, S.: Enhancing the formal foundations of BPMN by enterprise ontology. In: Albani, A., Barjis, J., Dietz, J.L.G. (eds.) CIAO!/EOMAS -2009. LNBIP, vol. 34, pp. 115–129. Springer, Heidelberg (2009). doi:10.1007/978-3-642-01915-9_9

11. Naplava, P., Pergl, R.: Empirical study of applying the DEMO method for improving BPMN process models in academic environment. In: 2015 IEEE 17th Conference on Business Informatics, vol. 2, pp. 18–26, July 2015

12. Figueira, C., Aveiro, D.: A new action rule syntax for DEmo MOdels based automatic workflow procEss geneRation (DEMOBAKER). In: Aveiro, D., Tribolet, J., Gouveia, D. (eds.) EEWC 2014. LNBIP, vol. 174, pp. 46–60. Springer, Cham (2014). doi:10.1007/978-3-319-06505-2_4

13. Heller, S.: Usage of DEMO methods for BPMN models creation. Master thesis, Czech Technical University in Prague. Computing and Information Centre (2016). https://ccmi.fit.cvut.cz/wp-content/uploads/2017/03/Heller_thesis_2016.pdf

DEMO Business Processes Design to Improve the Enterprise Business Continuity Plans

José Brás[1]([✉]) and Sérgio Guerreiro[2,3]

[1] Lusófona University, Campo Grande 376, 1749-024 Lisboa, Portugal
a21400334@alunos.ulusofona.pt
[2] IST, University of Lisbon, Av. Rovisco Pais 1, 1049-001 Lisboa, Portugal
sergio.guerreiro@tecnico.ulisboa.pt
[3] INESC-ID, Rua Alves Redol 9, 1000-029 Lisboa, Portugal

Abstract. Organizations are concerned in building resilience to mitigate the challenges that result from unpredictable and constant threat scenarios. Therefore, designing, implementing and operating solutions to increase the resilience is nowadays a key concern that must be addressed properly. The most usual resilience mechanism to deal with the unpredictable within the business ecosystem and correspondingly IT is to produce business continuity (BC) plans. On the one hand, a BC plan is usually established with the data collected from the operation and from the *as-is* holistic design of business processes (BP). However, BC plans are usually challenged by the insufficient, fragmented, inconsistent and incomplete information when capturing the enterprise' BP. On the other hand, the Business Impact Analysis (BIA) must be properly prepared to support all business activities and build good recovery strategies, demanding precise, concise, complete, coherent and consistent enterprise' business processes. This paper integrates and evaluates the DEMO holistic design of business processes with BIA. This solution contributes to the identification and enrichment of existing weaknesses in BC plans and thus improve resilience in case of threat. The validation of the solution is performed using an insurance company case study.

Keywords: Business continuity · Business transactions · DEMO · Business Impact Analysis

1 Introduction

In order to ensure an ability to operate on an ongoing basis and limit losses in the event of severe business disruption, companies need to have documented BC and Disaster Recovery (DR) plans. They must be reviewed on a periodic basis and updated to reflect changes in the business environment or within the supporting IT infrastructure. In general, the BC and DR plans need to identify critical functions, assets, processes and supporting systems in the business impact assessments, determine ways to operate key external services and internal functions in situations of disruptions, including alternative sites.

© Springer International Publishing AG 2017
D. Aveiro et al. (Eds.): EEWC 2017, LNBIP 284, pp. 99–107, 2017.
DOI: 10.1007/978-3-319-57955-9_8

After a disruptive situation, it is necessary to give an adequate response to the situations arising from it, and for this it is required to pre-establish the necessary measures to give an adequate response. The BC Plan allows the founding of strategies, procedures and critical actions necessary to respond and manage a crisis situation [16] and expresses an organization condition to responds to unexpected disasters, disruptions or sudden business changes [5]. These situations can result from a natural disaster, a catastrophe or might just occur from a simple accident and can cause the interruption of a service, a partial or total loss of the processes that sustain the business [9].

The British Standards Institution defines BC as the *"capability of the organization to continue the delivery of products or services at acceptable predefined levels following a disruptive event"*, also it defines business continuity management (BCM) as *"is a holistic management process that identifies potential threats to an organization and the impacts to business operations those threats, if realized, might cause, and which provides a framework for building organizational resilience with the capability for an effective response that safeguards the interests of its key stakeholders, reputation, brand and value-creating activities"* [1].

In a recent study [3], present a study on the adoption of DEMO (design methodology and engineering for organizations) as a complement to the process of analyzing the impact of processes on BC.

The adoption of DEMO in business ontology as become an important tool, not only as a form of process modeling, but also in the case of separations, reorganizations and post-merger integration of companies, in the form of process re-engineering. These facts are supported by several studies [6,14]. DEMO applications at a business level have demonstrated a great return in terms of modeling effort and in this particular case in this study, a systematic and reproducible abstraction capability that DEMO allows us to make and implement at the level of organizations, are beneficial to a business continuity function. The return on modeling effort (ROME) in this case is even higher by the integration of the two methodologies in the prevention and mitigation of risks [13].

The research problem is the following: *"the documentation to support a BC plan is often insufficient, fragmented and inconsistent, leading to misjudgment, misinterpretations and wrong impact calculations."*

This paper contributes with an integration between BC best practices and the DEMO concepts, to allow the construction of new models that enhance and serve as foundational knowledge to help building a BC Plan.

The rest of the paper is organized as follows. The next section of the paper, Sect. 2, introduces the ontological approach to BC combined with the concepts of DEMO theory and methodology, where the case study was founded. Afterwards, Sect. 3 presents conceptual foundations, particularly about BC concepts and the case study advances and the research methodology used in this paper. Subsequently, Sect. 4 presents the outcome learnings obtained from the case study and finally, Sect. 5 concludes and presents future work.

2   Background

Organizations now need to find solutions to face emerging challenges due to increasing complex organizational BP, consequence of the development of Industry 4.0 and by the Internet of Things [4,8,11,12]. An ongoing effort to develop new solutions to understand, implement, monitoring and continuous manage BCM [1] have motivated the industry, *e.g.*, the TOGAF 9.1 standard [15] that included the principle of BCP within the architecture principles framework.

Regarding Enterprise Ontology, nowadays it is fundamental to an organization not only to understand how the his business works but also to capture and retain knowledge. This can be done by modeling its BP with the use of an ontology for describing the elements, concepts, structures of the enterprise and the business itself. This knowledge can be captured and represented by modeling, here called the organizational model. Enterprise Ontology is a way of perceiving the construction and operation of a company independently of its realization and implementation. It is basically the highest-level constructional model of an enterprise, and the implementation model being the lowest one. It is also a way of gaining knowledge about how the organization works, allowing the development of a global awareness about the organization, as it allows the sharing of knowledge among individuals. This can be done through the representation of different organizational aspects, such as BP, resources (Technological artifacts, suppliers, key stakeholders etc.) and by representing the organizational structure.

Regarding Business Continuity Management (BCM), it allows a company to align its business to Governance, Risk and Compliance (GRC) frameworks. *"BCM Lifecycle shows the stages of activity that an organization moves through and repeats with the overall aim of improving organizational resilience. These stages are referred to as the Professional Practices and are made up of Management and Technical Practices"* [2]. The goal of the BIA is to detect and classify which business units/departments and processes are essential to the survival of the company and is the base of a BC and DR plans being a vital piece of the process in a comprehensive BC Program. Identifying correctly all BP will help evaluating the impact of disasters on business, providing the basis for investment in recovery strategies as well allowing to invest in prevention and mitigation strategies. After perform the BIA, the critical BP and dependencies are identified, which will allow the organization to prioritize resources and focus on the most critical processes first, when doing planning or actual BP recovery during an Severe Business Disruption. Finally, it is important to highlight that the International Organization for Standardization (ISO) Technical Committee (TC) 292, has already released the ISO/TS 22317:2015 - Societal security BCM Systems Guidelines for BIA [10]. Figure 1 shows and highlights the overall steps to accomplish the BIA process, along with the DEMO methodology. The outputs from the BIA and Risk Assessment (RA), which establishes and maintains the capability to resume business operations upon a disruption of service or event, will be the base and permit to develop and implement a resiliency strategy.

3 Organizational Requirements

In order to formulate the understanding of the applicability of DEMO and BC working together, to a BC strategy, it is necessary to provide important information to assist the understanding of the organization's needs and constraints. Companies that are regulated (telecommunications or bank and insurance, as an example) need to have consistent and well documented processes. They will form the basis of their activity which drives BC and DR planning, and need to be performed in a consistent manner and go through all relevant risks in order to create resilience. To do this, an organization must then identify organization's activities, functions, services, products, partnerships, supply chains, relationships with interested parties, and the potential impact related to a disruptive incident [7].

If we analyze Fig. 1, over steps (A to D) of the BIA process we can find links to the DEMO methodology where we find points of contact between the two methodologies with DEMO models, more specifically with the OCD model. This can serve as a base of analyses as these steps state that it is necessary to identify all activities that support the delivery of products and services, assess their impacts over time of not performing these activities, create an action plan by setting prioritized time-frames for resuming critical activities and to end, it is also essential to identify dependencies and all supporting resources for these activities, points where DEMO can give important contributions.

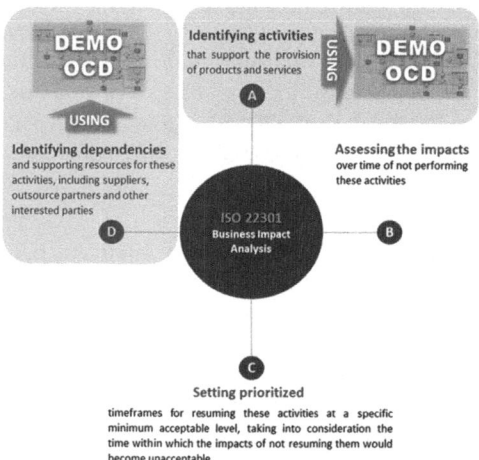

Fig. 1. Business Impact Analysis (Adapted from ISO 22301(2012)) [1, p. 15].

In summary, for a most comprehensive and credible BIA, all the important aspects of a process, interactions and dependencies, are required to correctly calculate the impact of a disruptive event on an organization. Since DEMO uses BP as the main focus of its methodology, its feature becomes an essential tool for studying or analyzing non-business impacts.

3.1 The Case Study Description

In addition to the investigation carried out by the literature review, a typical (potentially fictitious) BP of the insurance activity, involving different departments and external suppliers, was used as a case study. In this process, it was documented all the business flows, dependencies and activities between departments that were recaptured and documented using the DEMO methodology.

The BP and existing documentation (based on Rich Text Figures (RTF)) where analyzed and investigated, targeting all the major stakeholders that would be covered, and DEMO was used to re-evaluate, complete and re-validate the entire process, where applicable.

The method to capture a BP is normally a arduous assignment to accomplish. Stakeholders tend to use their natural "language" very much related to their specific area to describe their internal processes. Occasionally this implies misinterpretation by other departments that they interact with, since there isn't a common understanding about the way to capture business processes. Managers also need to easily audit and validate that the process complies with what is described at the business plan and check its completeness.

Methodologically, the original information which was collected, accessed and used during the course of the research, and originated the findings and final report of this paper. The collection and the analysis of data for this research was done doing meetings with process key stakeholders. They are the owners of the needed knowledge that will help in developing more in-depth understanding of the processes to be modeled. Also some semi-structured contacts on the subject with peers, was done and internal documents were also analyzed and reviewed.

4   Outcome Learnings

The research carried out under the DEMO integration with the BC process, more specifically in its conjunction with the BIA allowed to highlight some benefits by using the two methodologies together.

On the one hand, from the literature review, this research allowed the identification of a set of limitations that could be solved by the DEMO/BIA integration. These findings are presented in Table 1.

On the other hand, from the analysis of the conducted case study, the results were important to allow a better understanding of the problem and also to show the difficulties that the business continuity function finds and leads, either in the

Table 1. Limitations findings and DEMO/BIA solution from the literature review

Limitations	Benefit by using DEMO/BIA integration
Lack of representation of the external players role. This effect is highlighted when comparing the BIA with the OCD DEMO models	This limitation is well solved in Fig. 2 - Organization Construction Diagram (OCD), where all the actors involved in the business processes are fully expressed, including external actors together with the banks shared between them. These banks are not taken into account in a BIA study. Nevertheless they are mandatory to prepare a proper BC plan
Impact quantification for the business offered by the BIA, but not offered by DEMO	It is required in this case additional resources to other solutions, for instance: e3value, in which this case the BIA allows assess and collect evidences
Lack of detail and granularity of the BIA documents	Higher levels of granularity defining the DEMO compared to models and data models that are typically used by the BIA
Lack of state models used in the BIA calculation of risks and impacts on business. DEMO uses a formal state model to declare the rules, types of essential facts and relevant to real-world objects in its field of application	In DEMO, a state model is specified using an *"object-fact diagram"* (OFD) that is able to show the essential modelling of the world
The BIA does not show evidences of documentation where the dependencies between BP are sufficient to calculate the impact of disruptions into business processes	DEMO models highlight evidences of the dependencies of all the actors of the process. Offering a full detail of the conversation between actors within a business transaction

day-to-day basis, or to do the necessary anticipation of potential crises. These results are summarized in Table 2.

The use of a DEMO models will allows management to have a more broad and comprehensive view of all BP, permitting them to better assess the plans consistency, and to verify if it addresses all necessary activities to support BP. By using an OCD model it is easier to audit and confirm if the plan and the described processes are consistent and correct. Also permits to assess and audit whether the capture of the necessary resources and their interactions and dependencies have been properly carried out.

The combination of these two frameworks can allow an organization to have a more complete view of the structure of his business and also can serve to have a common language that can be more easily understood by all involved

Table 2. Limitations findings and DEMO/BIA solution from the case study

Limitations	Benefit by using DEMO/BIA integration
Lack of an inventory of external databases to the organization properly connected to the business processes to allow the BIA to estimate the impact of the loss of this information	DEMO help in qualifying and quantifying these data. With DEMO this is easier to detect because they are associated with the processes, this aspect reveals to be of great importance because it is an asset in terms of control and management of this information
Lack of a proper inventories of business rules and their mapping in the respective processes in a systematic way	It is evident by comparison with the existing documentation that DEMO models uses which are better organized - supported by the DEMO models (Action Model (AM) and Process Model(PM)) and its base methodology
Legal and Compliance related issues that arise due to often unauthorized and improper access to customer data, which can be done during and after recovery from disaster	These unauthorized access are incomplete documentation of reflection and failures and communication of rules and business standards in the documentation that supports the BP. In this case, the modelling allows using DEMO through OCD diagram (shown in Fig. 2 - DEMO OCD) which refers to the databases used in the process and AM and PM models support the following business rules and regulations
If a disaster occurs, the failures after a recovery must be mitigated. These can be derived from non-compliance with the legal aspects and regulation of the entities regulating the banking and insurance business. This needs to be easily identifiable and be aggregated to each processes description. One example can be the times required to respond in case of a complaint from a customer and its accessory penalties in case of default	DEMO allows a response to this need by using the model states to map all business process transactions in time and the AM and PM models for the rules and regulations

stakeholders. In addition to this it allows an easier way to redesign and re-engineering the business processes in the case of a major disruption by supporting the management board dealing with a crisis situation and fundamentally rethink how they will do their work.

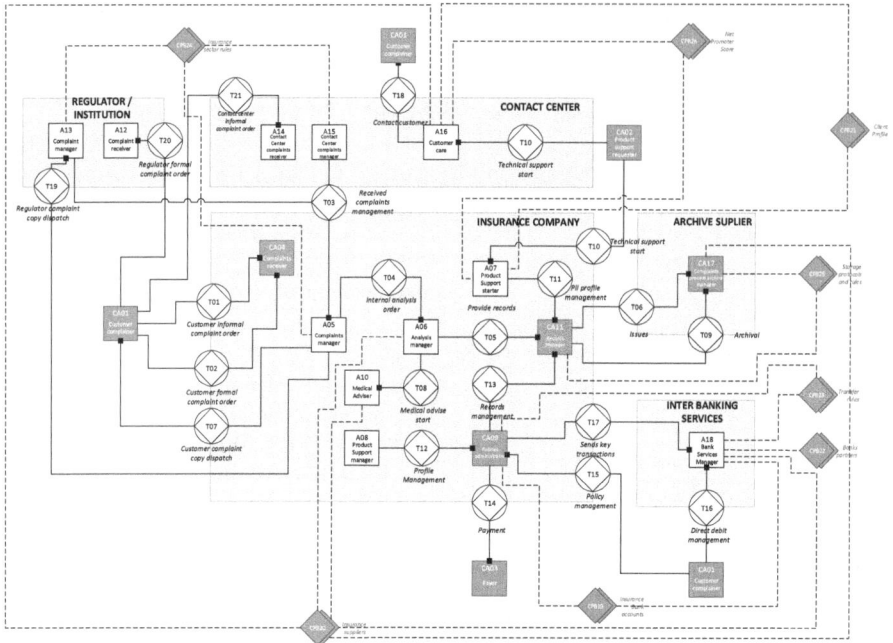

Fig. 2. DEMO OCD diagram - customer's service life cycle

5   Conclusions/Future Work

Regarding the benefits of the complementary use of DEMO with BIA, the present paper pinpoints that the DEMO methodology offers advantages to substantiate the BIA calculations of the impacts of a disruptive event on the business in more comprehensive and rigorous way by the use of DEMO models. The DEMO methodology used in the presented model allowed to represent more realistically all the main aspects and also all dependencies related to a BP. Although our initial results indicate positive benefits for combining the DEMO and BIA methodologies, a set of research questions require a more in-depth and methodical analysis of the results found. In particular, the role of DEMO models and their actual applicability to the benefit of their use with the BIA will in this case be further and more detailed. The possibility of using DEMO to re-engineer processes in the event of a real disaster occurring and some of the processes affected is one of the possible areas of interest and development of new studies and research.

Acknowledgments. This work was supported by national funds through Fundação para a Ciência e a Tecnologia (FCT) with reference UID/CEC/50021/2013.

References

1. ISO 22301:2012: Societal security – business continuity management systems – requirements. International Organization for Standardization (2012). https://www.iso.org/obp/ui/#iso:std:50038:en
2. BCI: What is business continuity? (2017). http://www.thebci.org/index.php/resources/what-is-business-continuity
3. Brás, J., Guerreiro, S.: Designing business continuity processes using DEMO. In: Pergl, R., Molhanec, M., Babkin, E., Fosso Wamba, S. (eds.) EOMAS 2016. LNBIP, vol. 272, pp. 154–171. Springer, Cham (2016). doi:10.1007/978-3-319-49454-8_11
4. Cerullo, V., Cerullo, M.J.: Business continuity planning: a comprehensive approach. Inf. Syst. Manag. **21**(3), 70–78 (2004)
5. COBIT: Cobit 5 for assurance. ISACA (2013). https://books.google.pt/books?id=FDdbAwAAQBAJ&lpg=PA1&dq=cobit%205&hl=pt-PT&pg=PA2#v=onepage&q=cobit%205&f=false
6. Dietz, J.L., Hoogervorst, J.A., Albani, A., Aveiro, D., Babkin, E., Barjis, J., Caetano, A., Huysmans, P., Iijima, J., van Kervel, S., et al.: The discipline of enterprise engineering. Int. J. Organisational Des. Eng. **3**(1), 86–114 (2013)
7. Drewitt, T.: A Manager's guide to ISO 22301: a practical guide to developing and implementing a business continuity management system. IT Governance Ltd. (2013)
8. Elliott, D., Swartz, E., Herbane, B.: Business Continuity Management 2e: A Crisis Management Approach. Routledge, London (2010)
9. Heng, G.M.: Managing Sustaining Your Business Continuity Management Program. GMH, Singapore (2007)
10. ISO: Societal security - business continuity management systems - guidelines for business impact analysis (BIA) (2015). http://www.iso.org/iso/catalogue_detail.htm?csnumber=50054
11. Krishnamurthy, T., Shetty, R.: 4G: Deployment Strategies and Operational Implications: Managing Critical Decisions in Deployment of 4G/LTE Networks and their Effects on Network Operations and Business. Expert's Voice in Networking. Apress, New York (2014). https://books.google.pt/books?id=-eCEBQAAQBAJ
12. Lasi, H., Fettke, P., Kemper, H.G., Feld, T., Hoffmann, M.: Industry 4.0. business. Inf. Syst. Eng. **6**(4), 239–242 (2014). http://dx.doi.org/10.1007/s12599-014-0334-4
13. Op't Land, M., Dietz, J.L.G.: Benefits of enterprise ontology in governing complex enterprise transformations. In: Albani, A., Aveiro, D., Barjis, J. (eds.) EEWC 2012. LNBIP, vol. 110, pp. 77–92. Springer, Heidelberg (2012). doi:10.1007/978-3-642-29903-2_6
14. Op't Land, M., Zwitzer, H., Ensink, P., Lebel, Q.: Towards a fast enterprise ontology based method for post merger integration. In: Proceedings of the 2009 ACM Symposium on Applied Computing, pp. 245–252. ACM (2009)
15. TOGAF: TOGAF Version 9.1. Open Group Standard (2011)
16. Tucker, E.: Business Continuity from Preparedness to Recovery: A Standards-Based Approach. Elsevier Science, Amsterdam (2014). https://books.google.pt/books?id=v95FBAAAQBAJ

Normalized Systems and Evolvability

Investigating the Evolvability of Financial Domain Models

Marjolein Deryck[1]($\boxtimes$), Ondrej Dvořák[2]($\boxtimes$), Peter De Bruyn[3]($\boxtimes$), and Jan Verelst[3]($\boxtimes$)

[1] Department of Accountancy and Finance, University of Antwerp, Antwerp, Belgium
marjolein.deryck@uantwerpen.be

[2] Faculty of Information Technology, Czech Technical University, Prague, Czech Republic
ondrej.dvorak@fit.cvut.cz

[3] Department of Management Information Systems, University of Antwerp, Antwerp, Belgium
{peter.debruyn,jan.verelst}@uantwerpen.be

Abstract. Evolvability is a characteristic dealing with change in Information Systems (IS). As the requirements evolve in time, the complexity of the system may increase. In turn, the ability to change it decreases. Consequently, the cost of a change can become unbearable. A domain model is an important abstraction covering key aspects of IS. Similarly to the IS it represents, it can suffer with the same evolvability issues. The goal of this paper is to assess combinatorial effects (CE) in a financial industry domain model, more specifically a domain model of financial risk management. It reveals difficulties related to identifying combinatorial effects in domain models in general and presents some insights on the nature of combinatorial effects on this level.

Keywords: Domain model · Normalized systems · Evolvability · Combinatorial effect

1 Introduction

The start of the 21st century is characterised by a digitalization of every aspect of society. The number of computers, internet access, applications and information increases exponentially and societal structures adapt to it [1–3]. At the same time, enabled by this digitalization, the expectations of customers and regulators compel enterprises to become more agile at every level. For Information Systems (IS), this can be interpreted as the capability to adapt to new functional requirements. These new functional requirements arise from changes in the environment and processes that surround the system as well as the user's experience [4]. This phenomenon is captured by Lehman's law of continuing change. At the same time Lehman posits the law of increasing complexity. It states that unless something is done to prevent it, the structure of an IS deteriorates over time [5]. The

© Springer International Publishing AG 2017
D. Aveiro et al. (Eds.): EEWC 2017, LNBIP 284, pp. 111–125, 2017.
DOI: 10.1007/978-3-319-57955-9_9

challenges linked to that are often explained as *technical debt* [6]. At the end, this results in the compulsory replacement of the existing system, which goes against the idea of evolvability, and cost effectiveness. Op't Land [7] attaches that a consequence of the deteriorating structure is an annual growth in budgets on development and maintenance. He clarifies that "Enterprises that decrease - or (even) keep constant - the IT budget will be faced with less satisfactory IT, decreased support of organizational changes, decreased business IT alignment and decreased situational awareness". Based on this statement, he identifies a challenge in a software development area - an approach that could help IT companies to develop software that exhibits high quality and is quickly continuously changeable over time. This challenge is captured in a term *evolvability*.

Following Cook's definition in [8], we altered it as:

Definition 1. *Evolvability is the capability of software products and their related design models to be evolved to continue to serve its customer in a cost effective way.*

A modular decomposition of a large system into smaller subsystems has long been identified as a way to improve evolvability and facilitate changes (e.g. [9–11]). To obtain its benefits, the subsystems should be partitioned in a precise, unambiguous, and complete way, and they should interact through standardised interfaces [11]. However, it is recognized that no universal measure or composition of a product's modularity exists ([12–14]). Different decompositions are possible and a firm should choose a decomposition that aligns with its objectives [14]. In their highly-referenced review paper, Campagnolo and Camuffo emphasise a lack of research on product design modularity that addresses market or industry specific factors possibly affecting product design modularity itself [14].

Normalized systems (NS) theory proposes a systematic methodology for a modular design with the objective of creating evolvable systems [15]. Its applicability as a theory for evolvable modular software systems has been proven by the development of critical software systems for multiple organizations [16]. The use of the theory in the broader scope of enterprise engineering has been demonstrated by research performed by De Bruyn, Huysmans and Van Nuffel [17–19]. However, the successful identification of Combinatorial Effects (CE) in some enterprise engineering instruments does not guarantee the general applicability. CE typically emerge at a very low and fine-grained level. Aggregating this in higher-level abstractions may hide the underlying impacts.

Moreover, the application of NS theory to specific industries only took off recently, e.g., [20]. Its mission is to demonstrate the factors that may hamper evolvable modular design. Such an analysis has not been done yet on the level of domain models, neither in the financial industry.

Thus, the purpose of this paper is to investigate the evolvability of domain models in the finance industry. We will focus on the sub-domain of market risk management, that is expected to be subject to regulatory changes in the coming years. Due to the importance of domain models in software development in general, the main focus is on analysing corresponding reference models using NS theory.

We present typical change requests, and we show their possible implementation. By doing so, CE inherent in the models are uncovered and described. The presence (or absence) of these effects indicate how hard (or not) it is for companies to implement changes within a reasonable time frame. Ergo, CE in this case may point to an increased risk for regulatory penalties if short-term changes are imposed.

In Sect. 2, we elaborate in general on what is meant by *a change*, and what might be its consequences. Next, in Sect. 3, we deeply introduce a finance domain model and we outline its possible changes. In Sect. 4 we revisit evolvability in domain models. We present related work in Sect. 5, and we conclude the paper in Sect. 6.

2 Evolvability

In the previous section we mentioned that the complexity of IS increases due to new functional requirements. However, in itself the requirement does not incline the increase of complexity. Rather, the corresponding changes effect it. Thus, in this section, we will clarify what it is meant by a change.

Clearly, IS can be changed at various levels. On the level of its source code, we usually refactor, optimise, add, or delete certain code constructs, e.g., functions in structured programming [21], or classes in object oriented programming [22]. On the level of a database, we alter, drop, or create new database objects, e.g., tables, triggers, and views in relational databases [23]. The modification of a software configuration is a change as well. Moreover, several cloud computing services offer a scalability option. For example, Microsoft Azure platform can adapt the system to an unexpected amount of workload by increasing or decreasing resources for an application [24]. Therefore, by introducing a new functional requirement "adopt to a workload automatically", we do not change the system itself at all. Yet the changes in the surrounding environment can affect it significantly.

Thus, in any kind of system, the formalisation of what a change means, is crucial. Below, we will show that NS formalises it by a term *task* as a subject to an independent change [15]. In structured programming, such a task is represented by a function. In Object-Oriented Programming (OOP), a method plays that role. A number of code lines usually implements the given function, respective method. These can be logically grouped into a sub-function, respective a sub-method, to signify they belong to a different change driver. Therefore, similarly to structured programming, or OOP, we have to formalise a change in the area of domain models, e.g., in the area of finance domain model.

2.1 Normalized Systems

The sections above describe the difficulty linked to the demand for evolvable systems. NS theory offers an answer to this challenge. It uses the formal foundations of system theoretic stability to study the transformation of (basic) functional requirements to the software primitives of a stable system [15]. This stable system is defined as "bounded input/bounded output" – if the system receives bounded input it should create a bounded output. It means that for a set of anticipated changes (i.e., changes in basic functional requirements) the impact on the system should only depend on the change itself and not on the size of the system [25].

NS proves that stable software systems can be created by the unification of four well known software design theorems. The *separation of concerns* principle states that in order to isolate change drivers an entity may only execute one task. Those tasks should furthermore exhibit *action version transparency*, meaning that a change in the task may not impact other tasks that call on the first task. Data used in tasks need to exhibit *data version transparency*. If data is modified this may not have an impact on the tasks that use the data. The final requirement is the call for *separation of states*, meaning that the status of every task should be kept.

The violation of either one of the four design theorems results in the creation of unwanted CE.

2.2 Combinatorial Effect

The combinatorial effect is defined in [15] as follows:

Definition 2. *... functional changes causing impacts that are dependent on the size of the system as well as the nature of the change...[are called] combinatorial effects ([25], p. 5).*

These unwanted effects stem from improper division in modules or an incorrect encapsulation of modules [15]. They may lead to large costs as changes will need to be implemented in multiple modules. This is what Mannaert et al. [15] call *the law of exponential ripple costs*.

However, modularity combinatorics might as well induce flexibility following the *law of exponential variation gains* [15]. It states that if an overall system consists of independent modules, the development and maintenance cost of those modules is the sum of all modules, whereas the number of variations is the product of all modules. In systems with multiple variants of the same unit of work, this leads to an exponential increase of possible combinations [15].

Modules following the NS theorems both leverage the opportunity described by the law of exponential variation gains while avoiding the unwanted CE.

3   Finance Domain Model

This paper covers the domain of risk management in financial institutions. Risk management has always been an essential activity of the banking sector, and since the financial crisis of 2008 it is under even closer scrutiny of local and international regulators [26]. Furthermore, banks themselves seek to optimise the internal models they use to calculate the regulatory capital, to avoid losses and capital punishment (the so-called plus-factor in case the actual loss exceeds the loss predicted by the internal model more than five times in one year) [27].

The Basel regulations discern three types of risk in the financial sector: credit risk (i.e., the risk that a counterparty will not honor his obligations), market risk (i.e., the negative financial impact of changing market conditions), and operational risk (e.g., fraud, settlement risk, etc.) [28]. Each of these risks cover multiple risk factors. For example, some factors that contribute to the market risk are changes in interest rates, share prices, commodity prices, inflation, foreign exchange rates, volatility and credit spread. The scope of this paper is the measurement of market risk using Value-at-Risk. The choice for this scope emanates first from the fact that this instrument shows clear disadvantages (i.e., lack of sub-additivity) and has been said to have played a significant role in the 2008 financial crisis. Second, eight out of the ten largest Belgian banks report the use of VaR in their annual (risk) report and the measure is accepted to calculate the regulatory capital for market risk. Therefore VaR might continue to play an important role in market risk management, but it will probably be subject to changes in the years to come.

The VaR is a single currency amount that reflects the maximal loss that is expected in the given time period. Regulators require at least 99% confidence on a ten day period, so a 10 day VaR(99%) of 500k means that the bank is 99% certain that the loss on the considered portfolio over the next 10 days will not exceed 500k. Note that VaR does not give any indication of the amplitude of the loss in case it is exceeded. The expected shortfall (ES), the calculation of which will be mandatory as from 2018, is adequate to that end [29].

3.1 Establishment of the Domain Model

The focus of this paper is a domain model of the market risk domain extended with a focus on market data import and trade repository. The model does not represent the situation in a single case company, but rather constitutes a realistic representation of common parts, based on the experience of the authors in multiple cases. The advantages of this approach are twofold. On one hand this generalization allows the abstraction of company-specific implementations that are not only the result of business requirements, but also of the company history, its specific systems and the quality of its implementation decisions. Even though the importance of these factors is recognized, they are not relevant in the light of this paper that aims to demonstrate the identification of combinatorial effects in reference models. On the other hand a real and detailed datamodel of a single case would require extensive access to the company's IS architecture, which might even not be readily available in the company.

3.2 Overview of the Domain Model

The domain model depicted in Fig. 1 abstracts the overall finance model. It displays three large parts. Situated in the upper left part (in gray) is a part related to the import of market data from an external market data supplier.

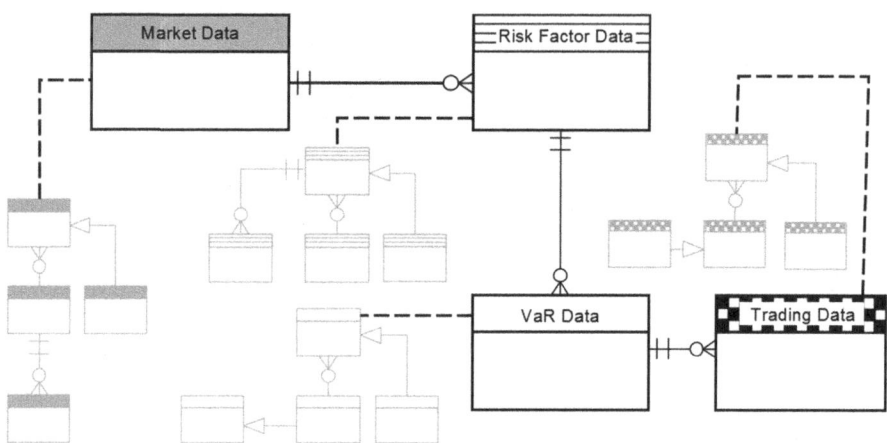

Fig. 1. Abstraction of a finance domain model

In the lower right part (with black cubes) a trade repository for foreign exchange and interest rate trades is depicted. The part in between relates to the calculation of Value-at-Risk following the historical method. The paragraph below explains how these blocks fit together in the process to calculate the VaR.

3.3 Business Process Introduction

Figure 2 schematically represents the VaR-calculation process with the use of the historical method. In this method the historical changes of the risk factors that have been observed during the last x days (often 300 to 500 days) are applied to the current trade portfolio. The process starts with the upload of relevant market information from external market data suppliers, such as Bloombergs, Reuters or others. The bank needs to specify which data should be downloaded at which moment. This is done by so-called schedulers. Certainly, also corresponding data-entities to store the information are needed. In the next step, the shift from one day to the other is calculated. This is nothing more than calculating the difference between yesterday's and today's value for, let's say, the 300 last days. Afterwards the one day shifts are scaled up with the factor $\sqrt{10}$ to obtain the ten day shifts necessary for the calculation of the ten day VaR. In the Full Reval Scenario the outstanding positions are valued against the ten day shifts. It means that for each outstanding position 300 possible profit and loss scenarios are calculated. The method is very simple, but it is heavy on calculating resources and available market data. For some deals with heavy pricing models it can be beneficial to calculate a proxy. This can be done with the use of sensitivities, e.g., delta, which reflects the change in value of a derivative when the value of the underlying changes. The sensitivities are used in the calculation of the profit and loss scenarios. Their VaR-calculations have their own parameters, including the alpha that indicates the certainty level. In Fig. 2 the method is

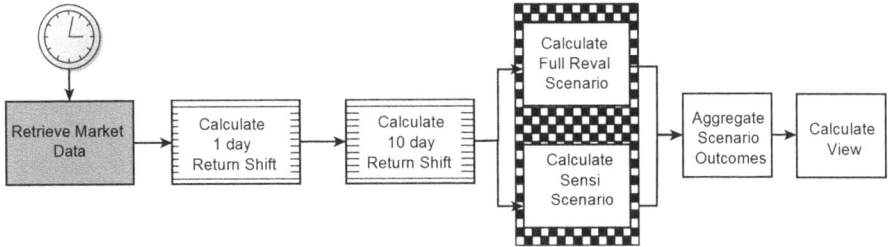

Fig. 2. High-level overview of the HVaR-process

represented by the rectangle below the full reval scenario-rectangle. They are situated in front of a background with black cubes, that represents the trades in the trade repository. The resulting outcomes of both the so-called full reval and sensi method need to be aggregated with the purpose of obtaining a single VaR-number in the end. As VaR is not sub-additive, this aggregation needs to follow strict business-rules determining the appropriate calculations for the appropriate positions. Afterwards, possible aggregated profit and loss outcomes are sorted from the most negative (i.e., loss) to the most positive. Based on the desired certainty level (usually 99%) the appropriate cut-off value corresponding to the alpha is selected as the VaR.

This section offered a high-level overview of the general VaR-process. The individual process phases are covered by the corresponding paragraphs in Sect. 4.

4   Revisiting Evolvability of Domain Models

The models described in Sect. 3 breaks down the VaR-calculation in blocks and classes needed to execute it. The different parts are linked with each other through interfaces. In short, the domain model exhibits a modular structure. As explained above, the specific scope of the model was chosen because of the expected regulatory changes in this domain. The characteristics of change and modularity are exactly two fundamental concepts in NS. Therefore, in the section below, we investigate the applicability of the theory on the VaR-domain model by applying some changes as defined below:

Definition 3. *[Changes to IS are] (1) the addition of new requirements; (2) the modification of existing requirements; and (3) the obsolescence of existing requirements ([15], p. 258).*

This results in changes in the domain model, e.g., adding or renaming an attribute, adding a relation, changing cardinality, etc. More specifically in this case, the appended changes are the addition of equity market data, a new version of 10 day VaR calculation, amendment of the alpha, and addition of a new product.

4.1   Revealing Combinatorial Effects

The proposed domain model is limited to the follow-up of the foreign exchange risk factor and the interest rate risk factor. The choice of the risk factors included in a model depends on the nature of the business conducted by the bank. However, these two are most commonly measured by the VaR. In this section, we will introduce a few changes to the model to investigate how the model reacts.

Addition of the equity risk factor. The first change is the addition of new risk factor, e.g., the equity risk factor. This means that the share prices need to be uploaded in the system. To this end a new data-entity *Share* needs to be introduced. Usually, this type of market data, along with information on bonds, futures, and funds, will inherit some general attributes from an *instrument* class. To induce the upload of share prices from an external market data supplier, the schedulers that start the retrieval of the data, need to be amended. This means an impact of the five schedulers that are currently identified. Moreover, the number of schedulers is not cast in stone itself. It is thinkable that new schedulers, e.g., quarterly or bi-yearly schedules, need to be introduced. That means that the introducing a new data-entity is not only dependent on the size of the system, but also grows along with the growth of the system. This demonstrates that the addition of a new data-entity of equity risk factor leads to combinatorial effects. Figure 3 schematically shows the discussed changes.

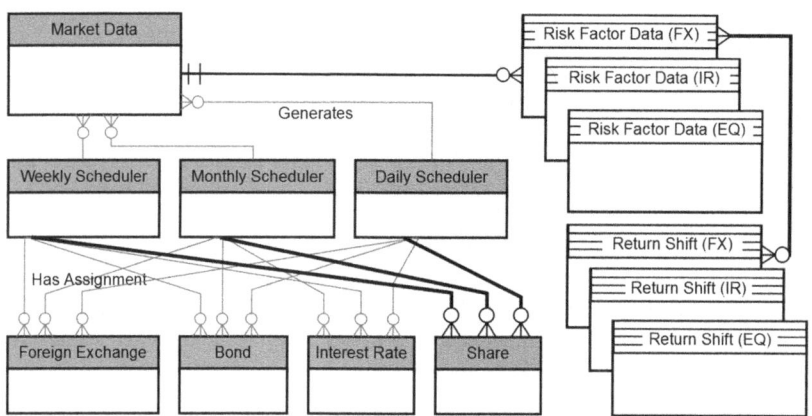

Fig. 3. Abstraction of a market data sub-model

Going further down the process, the market data are translated into Risk-FactorData (RFD). In the drafted domain model this is represented by a single RiskFactorData - entity. In fact, this hides two possible solutions – *generic entity*, or *multiversion entity*.

The generic entity is configured in a way that it can include all necessary details on forex, interest rate, and equity risk. It means that in this case a data-structure with superfluous data-fields are sent as an input for the return shifts.

This is an example of stamp coupling. Even though at first sight it may be tempting to tolerate this kind of coupling at the start, the risks associated with this structure are:

1. The overly large data structure using an extravagant amount of resources.
2. If one of the attributes changes or an additional attribute needs to be included, additional versions of the data-structure will need to be created. This leads to CE if updates need to be done on the different versions.

The multiversion entity solution helps to avoid this kind of coupling. We can create new versions of this Risk Factor Data (RFD) for each product. This would mean that at first there exists RFD for Foreign Exchange (FX), and RFD for Interest Rate (IR). We denote them RFD(FX) and RFD(IR) respectively. Upon the addition of the equity (EQ) risk factor, a new version, RFD(EQ), needs to be created. Going forward, when using separate versions of risk factor data, this logically leads to different versions of return shifts. An additional return shift would thus need to be created for equity risk. Figure 3 depicts the multiversion entity solution of RFD.

This shows that the addition of a new risk factor leads to multiple amendments in the system. To ascertain that these amendments are truly combinatorial effects, the number of amendments should even increase with the growth of the system. This is the case, which is demonstrated by the example below.

Amendment of 10 day VaR calculation method. In the current way of working we recognise two return shifts for each risk factor, i.e., the one day shift that is scaled to the ten day return shift. However, this kind of scaling will probably not be allowed anymore in new risk models. Regulators ask for a full calculation of the ten day VaR, and the adaptation of the existing domain models is unavoidable. This means that the return shifts in the new system will represent only one day changes. Therefore, next to the one day return shifts, a new data entity to capture ten day changes must be introduced. Such an adaptation must cover forex risk, interest rate risk, and equity risk. Conversely, if a new risk factor is added at this point in time, as before the risk factor data and the one day return shift need to be created. Furthermore, a 10 day return shift and possibly even more return shifts will be affected. This may happen if regulators estimate that the liquidity on the market has structurally changed and e.g., 1 month VaRs are necessary. If a new risk factor is added at this point in time, it requires more changes than the ones described for the addition of the equity risk factor.

The amendment of calculation methodology for +1 day VaRs shows another combinatorial effect when a full calculation (in contrast to the scaling calculation) is used. These changes are represented in Fig. 4.

At a basic level, the difference between two consecutive days can either be absolute or relative. In the scaling method, this difference can be multiplied with the square root of the number of days, both for the 1 day calculation (as $\sqrt{1} = 1$), as for the 10 day calculation (with $\delta_{10\,day} = \delta_{1\,day} * \sqrt{10}$).

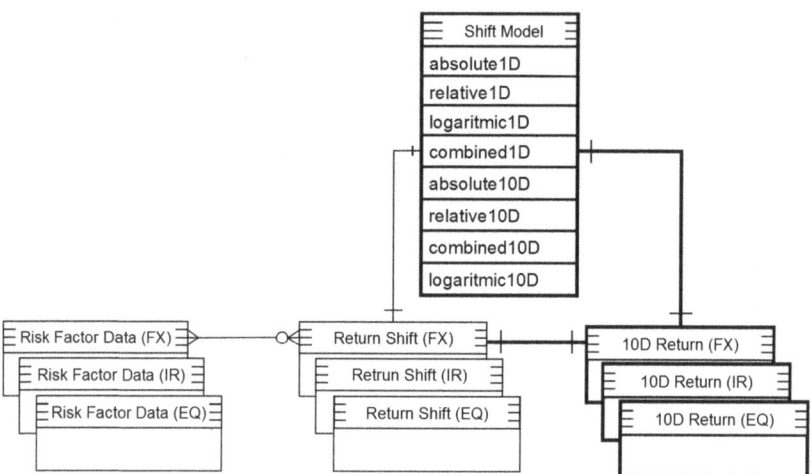

Fig. 4. Abstraction of return shift calculation sub-model

However, if a full calculation of a multiple day VaR is mandatory, this would mean that separate formulas for one day and ten day variations need to be created.

Unfortunately this is not the end of the story. Again, the impact of another VaR horizon does not always limit itself to the amendment of the two calculation methods. Although there are not an unlimited amount of possibilities, the number of calculation methods itself might increase as well. This consciousness emerged recently, with the prolonged low interest rates as an example. If the interest rate is at 0.01% at day 1, and rises to 0.02% at day 2, this is a relative change of +100%, and an absolute change of 0.01. Whereas the relative change exaggerates the impact, the absolute change would not show any differentiating power. A combination of relative and absolute elements in (one or more) 'mixed' calculation method could be implemented to remediate this. The introduction of logarithmic calculations could be envisaged as well. If these different methods are implemented, it means that a change in VaR horizon would be needed for each of them, hence demonstrating the definition of a combinatorial effect.

Addition of a new product. Another example is a change in the certainly level of VaR, which is the second important characteristic next to the time period under consideration. Currently the required alpha for the 10 day VaR is 1% maximum. However, if longer time-horizons are envisaged or in combination with other risk measures, regulators might be satisfied with a 2.5% alpha. Or conversely they might require a higher level of certainty by lowering alpha to 0.5% maximum. In the current domain model, this would mean that the alpha needs to be amended at two places: one time in the VaR-parameters (necessary for the full reval method), and one time for the calculation based on sensitivities (in the analysis parameters) (see Fig. 5). The choice between the two methods is implemented as a business rule at the Full Reval Scenario versus Sensi Scenario level.

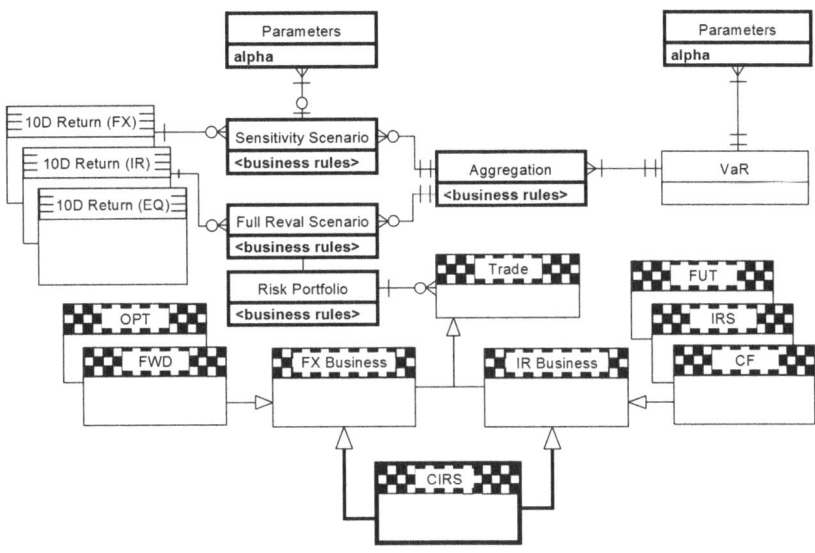

Fig. 5. Abstraction of model with new product type and changed alpha.

Now imagine yet another kind of change, e.g., the creation of a new type of product. In the current model only the clear-cut instruments from a risk factor point of view were described: two forex instruments (option and forward) and three interest rate instruments (swap, capfloor and future).

Let us see what happens if we add a currency interest rate swap (CIRS) to the portfolio. As the name suggests, this instrument contains both characteristics of the interest rate business (e.g., the respective interest rates of the currencies) and of the foreign exchange business (the exchange rate between the two legs). Therefore, a new data-entity will inherit some attributes from both businesses. Also, traders will need to be able to price this new instrument, so a new valuation model will be developed. This model will contain parts of IR-models and FX-models. As the CIRS contains IR-characteristics, the most probable scenario is the Full Reval Scenario, which means that the business rule that determines the calculation scenario (full reval or sensi) needs to be updated. Moreover, in this scenario a new Risk Portfolio needs to be created. This results in the adaptation of Aggregation Scenario, as it defines aggregation rules per portfolio. Again, as with the previous examples, it is the reaction of these changes to changes on another dimension. This clarifies whether these are true combinatorial effects or effects due to the original change. Imagine for example that the expression of change in foreign exchange rates does not happen in percentages anymore, but in pips (0.0001). It means that the original forex valuation models will need to be adapted. However, the CIRS model, that contains parts of a forex model, will be impacted as well. This applies to every new valuation model annexed with a new product. On the level of the business rules in the Full Reval/Sensi Scenario, the straightforward structure (with only one business rule) gets more and more complicated each time a new product is introduced (Fig. 5).

4.2 Insights in Exploring the Domain Model

NS aims to promote evolvability in large and complex systems. One of its key concepts is CE. In analyzing the domain model some insights on key characteristics of CE stemming from their purpose was provoked. To start with, CE are found in complex systems. Complex systems deal with entities that are prone to changes in different dimensions. Because of this, it seems difficult at first sight to uncover the CE in a domain model. Both knowledge on the domain to recognize the different change dimensions as insight on the reasoning to reveal CE are necessary. In turn, this implies that the use of CE to analyze small system that mainly consist of uni-dimensional constructs is less fit. Next, CE appear at a very basic level of analysis. Models, which are essentially abstractions of reality, may thus hide this complexity. This was shown in the examples above in which we identified CE on the level of business rules and attributes.

5 Related Work

Originally, the NS theory defines combinatorial effects on the software level. In subsequent research the concepts and theorems of the theory proved useful to evaluate evolvability on the enterprise level and in process models [17–19]. Eesaar [30,31] focuses on the database level of IS, and tackles the common ground between database normalization and NS. His research concludes that even though both theories overlap at some extent in their goal (avoid update anomalies), manner (multistep nonloss-decomposition), and result (an increased number of smaller tables/modules), NS covers a broader field with its focus on combinatorial effects [31]. Domain models are frequently used as the input for the creation of data models, and they often display the desired normalization form of the final data model.

In line with the conclusion of Eesaar [30] the current paper illustrates that normalization is a first, but by no means sufficient, step to avoid combinatorial effects.

The related work in the area of risk management does not seem to provide relevant research on evolvability and implementation. Implementation in this domain focuses mainly on the choice of unbiased parameters and financial calculations, which is not the scope of our research.

6 Conclusion

This paper started by establishing the importance of evolvability for current IS. IS are built to fulfil certain requirements, but these requirements tend to change over time. NS theory proposes four design theorems to create evolvable software, defined as software free from combinatorial effects. A combinatorial effect on this level is the impact from a change that does not only depend on the change itself, but also depends on the size of the system. CE can also be identified on the level of process models, and guidelines exist to avoid them at this level [17].

However, when applying the concept of CE on business models, we need to take into account that a high level of abstraction may hide CE present at lower levels. Therefore the research for CE typically needs to happen on a low level of aggregation. The current paper investigates whether the concept of combinatorial effects can be applied on the level of domain models, which represent an abstraction of the IS. And if this is the case, what can be concluded with regards to the evolvability of domain models. By investigating a partial domain model of financial risk management, it is demonstrated that combinatorial effects do exist in this domain. Earlier research revealed the existence of CE in, e.g., accounting systems, education programs, and ERP-systems [20,32–34]. It supports the belief that CE exist in more, if not all, economic sectors and types of IS.

It also demonstrates that the application of the concept of CE may not always be as straightforward as one might be led to believe by the simplicity of its definition. In practice, it is not always clear which effects are 'proprietary' effects of the change itself, and which ones are due to the size of the system. The reasoning build up in the examples of this paper demonstrates a possible way to proceed. It leans on the definition of a combinatorial effect as being dependent on the size of the system, hence, its amplitude needs to grow when the system grows. The system grows by adding changes of another dimension. This is demonstrated in the paper by considering different scenarios for one change. This way of working emphasizes the multidimensional nature of CE. Hence, it also contributes to the understanding of the difference with database normalization, which removes some data redundancy but in a one-dimensional way. In fact, only the CE emanating from the definition of the alpha at two different places would be solved by simply applying database normalization rules and isolating it in a new entity.

The changes that were applied on the original domain model show that CE manifest themselves on multiple levels. In the most visible form a change leads to the amendment of the domain model by the necessity to add additional classes (see change of 10 day VaR calculation) or multiple relations (2 inheritance relations for CIRS). In other cases, the effect is situated on the level of the attributes of (a) class(es). The increase in shift methods and the adjustment of the alphas are examples of this. In yet other cases the business rules linked to certain classes need to be amended. This is shown by the addition of the new product type CIRS, which alters business rules in multiple classes.

Because of this, future research should formalize the definition of 'change' at domain model level. Moreover, future research should offer recommendations to avoid the described CE. Moreover, the current paper is strictly limited to a well defined scope that reflects common practices in the sector, but nevertheless disregards others. Therefore, the future research should broaden the scope of current paper. The model should be expanded to encompass the entire scope of market risk, including multiple risk factors, other VaR-calculation methodologies, backtesing and stress testing practices, and future requirements such as expected shortfall calculation. Moreover, as we know now that CE do exist in one sub-part of the financial industry, their existence in other sub-parts may be investigated. Also, these sub-parts can be considered as modules in themselves. The information passing from one part to the other acts as the interface. Therefore future research could focus on possible CE at a higher level. Of course, difficulties regarding the hiding of CE at higher levels of abstraction need to be taken into account.

The current paper focusses mainly on data requirements in the domain. Future work could consider the same domain from a dynamic point of view and focus on e.g., process models like the HVaR process in Fig. 2.

Acknowledgement. We warmly thank Belfius and COPS GmbH which enabled us to settle down the discussed financial domain model. This research has been supported by SGS grant No. OHK3-006/17.

References

1. Hilbert, M., López, P.: The world's technological capacity to store, communicate, and compute information. Science **332**(6025), 60–65 (2011)
2. Marketline: Mobile apps in the united states. website, February 2016. https://store.marketline.com/report/mlohme7979--mobile-apps-in-the-united-states/
3. Fuchs, C.: Internet and Society. Social Theory in the Information Age. Routledge, London (2008)
4. Ciraci, S., Van Den Broek, P.: Evolvability as a quality attribute of software architectures (2006)
5. Lehman, M.M.: Programs, life cycles, and laws of software evolution. Proc. IEEE **68**(9), 1060–1076 (1980)
6. Behutiye, W.N., Rodrguez, P., Oivo, M., Tosun, A.: Analyzing the concept of technical debt in the context of agile software development: a systematic literature review. Inf. Softw. Technol. **82**, 139–158 (2017)
7. Op't Land, M., Krouwel, M.R., Dipten, E., Verelst, J.: Exploring normalized systems potential for Dutch MoD's agility. In: Harmsen, F., Grahlmann, K., Proper, E. (eds.) PRET 2011. LNBIP, vol. 89, pp. 110–121. Springer, Heidelberg (2011). doi:10.1007/978-3-642-23388-3_5
8. Cook, S., Ji, H., Harrison, R.: Software evolution and software evolvability. Technical report, University of Reading, UK (2000)
9. Simon, H.: The architecture of complexity. Proc. Am. Philos. Soc. **106**, 467–482 (1962)
10. Sanchez, R., Mahoney, J.T.: Modularity, flexibility, and knowledge management in product and organization design. Strateg. Manag. J. **17**, 63–76 (1996)
11. Baldwin, C., Clark, K.: Managing in an age of modularity. Harv. Bus. Rev. **75**, 84–93 (1997)
12. Gershenson, J.K., Prasad, G.J., Zhang, Y.: Product modularity: definitions and benefits. J. Eng. Des. **14**(3), 295–313 (2003)
13. Sako, M.: Modularity and outsourcing. In: The Business of Systems Integration, pp. 229–253. Oxford University Press (2003)
14. Campagnolo, D., Camuffo, A.: The concept of modularity in management studies: a literature review. Int. J. Manag. Rev. **12**, 259–283 (2010)
15. Mannaert, H., Verelst, J., De Bruyn, P.: Normalized systems software architectures. Normalized Systems Institute (2016)

16. Huysmans, P., Verelst, J., Oost, H.M.A.: Integrating information systems using normalized systems theory: four case studies. In: 17th IEEE Conference on Business Informatics, 13–16 July 2015, Lisbon, Portugal. IEEE (2015)
17. Van Nuffel, D.: Towards desdesign modular and evolvable business processes. Ph.D. thesis, University of Antwerp (2011)
18. Huysmans, P.: On the feasibility of Normalized Enterprises: applying Normalized Systems Theory to the high-level design of enterprises. Ph.D. thesis, University of Antwerp (2011)
19. De Bruyn, P.: Generalizing normalized systems theory: towards a foundational theory for enterprise engineering. Ph.D. thesis (2014)
20. Vanhoof, E.: Evolvable accounting information systems: applying design science methodology and Normalized Systems theory to tackle combinatorial effects of multiple GAAP. Ph.D. thesis, University of Antwerp (2016)
21. Dahl, O.J., Dijkstra, E.W., Hoare, C.A.R.: Structured Programming. Academic Press Ltd., Cambridge (1972)
22. Smith, B.: Object-oriented programming. In: Advanced ActionScript 3, pp. 1–23. Springer (2015)
23. Codd, E.F.: Relational database: a practical foundation for productivity. Commun. ACM **25**(2), 109–117 (1982)
24. Webber-Cross, G.: Learning Microsoft Azure. Packt Publishing Ltd., Birmingham (2014)
25. Mannaert, H., Verelst, J., Ven, K.: The transformation of requirements into software primitives: studying evolvability based on systems theoretic stability. Sci. Comput. Program. **76**(12), 1210–1222 (2011)
26. Saunders, A., Cornett, M.: Financial Institutions Management. A Risk Management Approach, 7th edn. McGraw-Hill, New York (2011)
27. Basel Committee on Banking Supervision: Minimum capital requirements for market risk. website, January 2016. https://www.bis.org/bcbs/publ/d352.pdf
28. Basel Committee on Banking Supervision: International convergence of capital measurement and capital standards. website, June 2006. http://www.bis.org/publ/bcbs128.pdf
29. Basel Committee on Banking Supervision: Fundamental review of the trading book. website, May 2012. http://www.bis.org/publ/bcbs219.pdf
30. Eesaar, E.: The database normalization theory and the theory of normalized systems: finding a common ground. Baltic J. Modern Comput. **4**(1), 5–33 (2016)
31. Eesaar, E.: On applying normalized systems theory to the business architectures of information systems. Baltic J. Modern Comput. **2**(3), 132–149 (2014)
32. Oorts, G., Mannaert, H., De Bruyn, P., Franquet, I.: On the evolvable and traceable design of (Under) graduate education programs. In: Aveiro, D., Pergl, R., Gouveia, D. (eds.) EEWC 2016. LNBIP, vol. 252, pp. 86–100. Springer, Cham (2016). doi:10.1007/978-3-319-39567-8_6
33. Chongsombut, O., Verelst, J., De Bruyn, P., Mannaert, H., Huysmans, P.: Towards applying normalized systems theory to create evolvable enterprise resource planning software: a case study. In: Lavazza, L. (ed.) The Eleventh International Conference on Software Engineering Advances: ICSEA 2016, 21–25 August 2016, Rome, Italy, pp. 172–177 (2016)
34. De Bruyn, P., Van Nuffel, D., Verelst, J., Mannaert, H.: Towards applying normalized systems theory implications to enterprise process reference models. In: Albani, A., Aveiro, D., Barjis, J. (eds.) EEWC 2012. LNBIP, vol. 110, pp. 31–45. Springer, Heidelberg (2012). doi:10.1007/978-3-642-29903-2_3

Exploring Design Aspects of Modular and Evolvable Document Management

Gilles Oorts$^{(\boxtimes)}$, Herwig Mannaert, and Peter De Bruyn

Normalized Systems Institute, University of Antwerp, Antwerp, Belgium
{gilles.oorts,herwig.mannaert,peter.debruyn}@uantwerp.be

Abstract. Over the past decades, technological advances have drastically changed the way organizations work. One thing that has not changed however, is that they are still required to draft, maintain and manage documents. Nowadays most of these documents are drawn up and stored in an electronic way. Yet their structure is in essence still the same as their analogue and physical predecessors. In this paper, we present a new modular approach to document management that enables the design of content-agnostic and evolvable documents. This new approach imagines documents as multidimensional and ever-changing entities instead of mere static representation of their analogue predecessors. Based on modularity and Normalized Systems reasoning, this approach represents an alternative to the out-of-date paradigm of document management that is currently used. Modularity-based document management leads to easier maintenance of the text modules as they offer a clear aggregate structure and any information is stored in only one text module. This enables both re-use and greater versatility of the information stored in the text modules. As such, the approach presented in this paper enables the creation of truly evolvable documents according to the Normalized Systems theory.

Keywords: Normalized Systems theory · Modularity · Document management · Evolvable documents · Modular documents · Text modules

1 Introduction

Over the past decades, technological advances have drastically changed the way organizations work. One thing that has not changed however, is that they are still required to draft, maintain and manage documents. These documents are present in organizations in an abundance of forms, such as books, spreadsheets, slide decks, manuals, legal contracts, emails, reports, etcetera. Nowadays most of these documents are drawn up and stored in an electronic way. Yet their structure is in essence still the same as their analogue and physical predecessors. Take for example electronic invoices. They often look just the same as their equivalent paper version. Therefore they are frequently just printed by the receiving organization and filed on paper. Similarly, the safety procedures in a manufacturing

© Springer International Publishing AG 2017
D. Aveiro et al. (Eds.): EEWC 2017, LNBIP 284, pp. 126–140, 2017.
DOI: 10.1007/978-3-319-57955-9_10

plant are often just saved as a pdf-file on a server instead of being printed and handed out (as was the case in the old days).

Lots of opportunities to optimize and rethink document management were missed during the revolution in Information Technologies (IT). Because of the lack of a theoretical approach to document management, most efforts in organizations were limited to just digitizing documents, i.e., transforming them from analogue to digital form as monolithic blocks. In this paper, we present a new modular approach to document management that enables the design of content-agnostic and evolvable documents. This new approach imagines documents as multidimensional and ever-changing entities instead of mere static representation of their analogue predecessors. Based on modularity and Normalized Systems reasoning, this approach represents an alternative to the out-of-date paradigm of document management that is currently used.

In Sect. 2, we will first demonstrate the need for variability and evolvability in documents. Next, we will show how to achieve these document characteristics using the principles of modularity and evolvability based on Normalized Systems theory in Sect. 3. Based on these insights, we discuss how current approaches address some aspects of document evolvability in Sect. 4. This brings us to the principal part of this paper, in which we give an overview of the most important aspects in creating modular and evolvable documents (Sect. 5) and how these can be implemented in practice (Sect. 6).

2   The Need for Evolvability and Variability of Documents

Documents need to incorporate changes and evolve over time. Information has to be added, changed or removed from organizational documents on a regular basis. In the current ever-changing business environment, companies need to be able to adapt to changing requirements of customers, government, competitors, suppliers, substitute products or services, and newcomers to the market [10]. These influences require adaptations in the supporting systems of the company, including documents. In most organizations nowadays, these documents are managed in a digital way. Therefore they can be more easily edited (by multiple people) throughout time, will be changed more frequently and have several concurrent variants.

In today's agile markets, countless *change events* require changes to the supporting document over time. Consider for instance the following events:

- a new legislation may require companies to add additional checks and safeguards to their documented corporate governance guidelines in order to avoid corporate fraud;
- a smartphone manual may need to be updated because a new version with added functionality was designed and is put into production;
- a financial report may need to be updated with new information and numbers of a new financial quarter;

These are just some examples of business changes that require adaptations in corporate documents. Listing all change events that require documentation changes would be impossible, as there is an infinite amount of events that require changes to the documents. To cope with this vast amount of change events, documents need to be *designed* to be changed with ease from the start and be *evolvable*. This will be discussed in the next sections of this paper.

The continuous change of documents also causes *variability* in documents. Adaptations in documents may produce several versions of documents as all these versions might need to be preserved. Consider for instance the following possible variants [7]:

- a similar slide deck on a subject may be created for a one day seminar to a management audience, a one week course for developers, a full-fledged course for undergraduate students;
- a product manual may be drafted in different languages, several product variants (standard – professional – deluxe) may contain a partly overlapping set of production parts requiring similar yet different manuals, etcetera;
- similar, but slightly different, legal documents (contracts) may be drafted for different clients purchasing the same service (based on the same contract template), etcetera;

These are once more just a few examples of how different versions of a document may arise, as the potential cases resulting in multiple versions of a document are limitless. To manage concurrent and successive document variants, most companies use so-called Document Management Systems (DMS). The problem with (most of) these systems is that they store documents at the "document" level. This corresponds with considering documents at their most aggregate level of modularity (i.e., the complete document as a single entity). As we will discuss in this paper, we propose a solution to store and manage documents at more fine-grained modular levels, enabling the creation of evolvable and reusable documents.

3   Modular and Evolvable Documents

3.1   Modular Structure of Evolvable Documents

Using the concept of modularity as a principle to design systems has proven to be very successful in various domains. It has for example been cited to be useful in product, system and organizational design [1,2].

Based on these understandings, previous research has shown the benefits of modularity in systems such as accountancy, business processes and enterprises. This research has proven that all these systems can be regarded to be modular [3,11,12]. Furthermore the scholars have shown that applying the modularity principle to systems benefits the design, maintenance and support of the system.

We are convinced that documents can also be considered to be clear examples of modular structures. Take for instance these examples [7]:

- A book or a report typically consists of a set of *chapters*. Each of these chapters will contain a set of *sections*, subsections, subsubsections, and so on. Each of these (sub)sections can then contain *paragraphs* with the actual text, tables and/or figures;
- A product manual will contain guidance *sections* regarding the different product parts and/or functionality;
- A legal document may contain different *parts*, within each part different *clauses*, and each clause may contain different *paragraphs*.

All of these document parts can be considered to be text modules. In our approach, we define a module as a part of the system that is used or activated separately. Once a part of the system cannot be used or activated as such, it is considered to be on a sub-modular level. Applying this general definition to documents, we can define a document to be a modular system. As such, a document consists of a set of *text modules*. A text module can subsequently be defined as a part of a document that can be included or referred to separately in a document.

Text modules should be identified and defined based on change drivers. The level of granularity (i.e., what information to combine or separate in text modules) should be based on whether the information is linked to one or multiple change drivers. A change driver can trigger some text parts to need to be changed in a document. As such, text parts linked to different change drivers should be broken up into separate modules. In this way, a change driver only requires changes to be made in a single module of a document.

To obtain documents that easily change over time, using the concept of modularity to define modular documents is just a prerequisite. For documents to easily incorporate changes, they need to exhibit evolvability. The *Normalized Systems (NS) theory* was proposed to achieve such evolvability, based on the modularity concept. Although the theory was originally defined for software, its applicability and value in other domains (e.g., organizational design, business processes, accountancy) quickly became clear [3,11,12].

To design evolvable systems, it is fundamental to eliminate so-called *combinatorial effects* according to NS theory. These effects occur when the impact of a change to a system depends on the size of the system they are applied to [8]. As such, the impact of the change would not solely depend on the nature of change itself. Assuming systems become ever more complex over time, combinatorial effects would become ever bigger barriers to change. Therefore, a system designed to not include any combinatorial effects would require less effort to be adapted when change requirements occur and thus be (more) evolvable.

To obtain evolvability, NS theory proposes four *theorems*, two of which are of importance in this paper [8]:

- *Separation of Concerns*, stating that each change driver (concern) should be separated from other concerns. This closely relates to the concept of cohesion;
- *Version Transparency*, stating that modules should be updatable without impacting any linked modules;

In practice, the consistent application of these theorems results in a very fine-grained modular structure, as will be shown in this paper.

3.2 Exploring Cross-Cutting Concerns

An important concept defined in NS theory is *cross-cutting concerns*. This concept is often used in information technology and refers to functionality or concerns that cut right across the functional structure of a system. Similarly, in the context of documents the concept refers to functionality or concerns that (may) matter to or affect several aspects of a document. According to the Separation of Concerns principle, cross-cutting concerns should be encapsulated. As such, changes can be made to the cross-cutting concerns without impacting the whole document. As we will illustrate in this paper, this is not self-evident as the functionality of these concerns are embedded deep down within documents.

The implementation of cross-cutting concerns should be available in a way that they do not depend on each other or on the internals of the base text modules. A text module should, for instance, be able to refer to another text module (i.e., use a specific reference system) without this having an impact to the typesetting/layout of the aggregate document.

Concerning documents, two types of cross-cutting concerns can be discerned. These are the concerns resulting from (1) the nature of documents and (2) the content of the documents (i.e., the underlying artifacts).

The *documentation cross-cutting concerns* are generic concerns that occur in any document. These concerns do not dependent on the content of the document, but need to be provided for all documents.

An important cross-cutting concern of this type is a mechanism for "relative" embedding of text parts in the hierarchical structure of documents. This means one should be able to include a text module on several hierarchical levels in a document without this inclusion causing any changes in the text module. As such, a text module can be variably used as a chapter, section, subsection, etc. without any changes to the module.

Another cross-cutting concern in the context of documents might be a reference mechanism to allow the author to refer to other sections, figures or tables within the overarching document (but possibly contained within other text modules), bibliographic items (e.g., another book or paper), etcetera. Preliminary research shows there are several other cross-cutting concerns for documents, such as for example:

- typesetting (layout),
- language,
- target audience,
- etc.

In addition to the documentation cross-cutting concerns, there is a second type of concerns that is specific to the underlying artifact(s) described in the document. We call these *content cross-cutting concerns*. They originate from the content or descriptions of the artifact(s). Take for example technical data sheets included with a bucket of paint. These documents will contain preparation instructions, application instructions, safety guidelines, storage guidelines, mixing directions, etc. These are necessary subjects needed in the description of every paint and can, as such, be defined as content cross-cutting concerns.

4   Attempts in Modular and Evolvable Document Management

Based on the requirements for modular and evolvable documents described in the previous sections, the authors studied some available document systems that showed promise in providing some of the requirements. One of the best systems in which the researchers could recognize some of the modularity and evolvability aspects mentioned is the LaTeX document preparation system, which is widely used for the formatting of scientific documents and was already available in the 1980s. A significant advantage of LaTeX in term of modularity is that it allows to include files into other files over multiple levels. This enables the aggregation of several text modules and the re-use of these modules into multiple versions or variants of overarching documents. When a base text module—which is included within several aggregate documents—is changed, this change can be automatically incorporated in all aggregate documents. As such, the user will be required to make the change only once and all relevant documents can be updated. The LaTeX system uses .tex files, which are text files containing LaTeX-specific commands. These can be build into different output formats (e.g., pdf or ps) and –via certain extensions– to .doc and simple HTML files by the LaTeX document preparation system. Moreover, the same base text modules can be used to build files according to different layouts by using document classes (e.g., book, paper) and stylesheets (e.g., applying the organizational house style to correspondence). The LaTeX system also allows to reference tables, figures or sections in a document in a relative way. This means references are generated during the build of the document and things can be referenced no matter where they are included in the final document.

However, the LaTeX system also has some important limitations in terms of evolvability. First, the actual hierarchical structure of a document (i.e., whether something is a chapter, section, subsection, subsubsection, etcetera) should be hard coded within .tex files. This severely limits the potential for base text modules to be freely combined into aggregate documents, as it can only be incorporated in other documents on the same document level. Second, the inclusion mechanism of individual LaTeX text files within other files does not force users to have a prescribed modular structure (e.g., to have separate text files for separate sections). As such, determining the modular structure of the text modules and aggregate documents is still left up to the determination of the user. Finally, LaTeX does not seem to automatically embed all the documentation cross-cutting concerns (e.g., a set of text property options which are used to automatically select the right versions of certain text excerpts). However, the systems does allow for extensions upon its basic functionality through the use of (predefined) macros. Therefore, we believe that this (or a similar) text preparation systems is an interesting and suitable candidate to explore the optimization of the modular structure of documents.

Another text formatting syntax that has gained popularity over the past decade is Markdown [4]. Originally created as text-to-HTML conversion tool, it has been expanded upon with document conversion tools such as Pandoc [5]. This allows documents to be transformed from about 15 source formats into even more output formats. Pandoc can merge files as a part of this transformation, so you can easily render multiple files into a single output. Markdown allows the user to easily mark layout with punctuation characters (e.g., an * symbol for emphasis, a = or - symbol to underline some words and a certain amount of # symbols to define the header level). Although this allows for a evolvable implementation of the typesetting cross-cutting concern, the Markdown/Pandoc system shows some serious limitations in terms of evolvability. Similar to the LaTeX solution, Markdown/Pandoc requires hard coded document structure and does not force a modular structure. It clearly also does not support all cross-cutting concerns.

In the next sections, we will examine the requirements for evolvable documents and if they can be accommodated with these readily available systems.

5   Design Aspects of Modular and Evolvable Documents

5.1   Coupling and Ripple Effects

Content within documents is often characterized by a high degree of dependencies. This kind of coupling is unavoidable. A well written and coherent document will build upon the knowledge or concepts explained in previous sections or chapters. In the case of this paper for example, Sect. 3 requires it should always be included *after* Sect. 2 as the latter requires the reader to have read and understood the former. In legislation, certain statutes require the existence of other statutes in a law. Removing or adapting particular statutes in the law will therefore cause other parts of the law to become inconsistent or change their meaning inadvertently. Although well-designed texts might reduce coupling to a certain extent by creating clearly cohesive and somewhat self-embedded document parts, not all coupling can be avoided.

However, other types of coupling can be included in documents that are not related to its content or buildup. These avoidable dependencies are positively detrimental for evolvability and variant creation. Often this coupling is caused by managing documents at an aggregate level instead of at a lower text module level. This means some content that is embedded in multiple documents is duplicated in a hard coded way within each of these documents. Furthermore, most document systems also combine multiple cross-cutting concerns (e.g., content, output type, layout, etc.) in a single document file/module. Consider for instance the following type of changes and their impacts [7]:

- The organization documents are drafted for, changes its house style. Often, this will require the manual adaptation of all versions and variants of all reports and slide decks (and all its individual slides) in order to be consistent with this new house style.
- A product part is used within several end products. Therefore, the text explaining the working of the concerning product part is embedded (here: duplicated) within all user manuals for all end products which contain this product part. The manuals for the end products are available in hard copy, an electronic offline document (e.g., pdf or .doc) and an online wiki (HTML pages). Moreover, as the product is sold internationally, versions of the user manuals have been created for all official languages spoken by the end product users in all output formats. When a new bug has been found for a product part which should be mentioned to the end users, this implies a change to all manuals of all end products which contain this product part, in all relevant languages and all output formats.
- After in-depth research, a researcher discovers that his definition of a particular concept needs to be adapted or refined. Typically, this will require him to scan through all papers, reports and slide decks (as well as all their possible versions and variants related to target audiences, output formats, languages, etcetera) to identify and adapt this concept definition.
- After some further research, one finds out that an additional figure and an additional paragraph might be useful within the sequence of steps generally taken to explain a certain reasoning. Applying this additional step would have an impact on all slide decks, reports, etcetera and all their versions, variants, etcetera.
- An organization decides that from now on, all legal contracts should mention that in case of disputes, trials will be held at a Belgian court and Belgian legislation is applicable. Typically, this will require the organization to check (and if the relevant clause is not present, adapt) all of its active legal contracts.

These examples show the impact of combinatorial effects in documents. By managing documents at a text module level, some of these detrimental effects could be avoided. This modular approach would also allow various types of document aggregations without the duplication of content in every document a text module is re-used.

5.2 Version Control

Currently, as mentioned above, most Document Management Systems manage documents as one single monolithic unit. Classic word processors or presentation software mostly work in the same way; they only allow the creation and editing of documents as one monolithic whole. This approach is also applied in their management of versions and variants, as they are also managed at the document level: including the same product part description in multiple product manuals via relative references instead of copy-paste is then, for instance, impossible. As a result, the different variants or versions of a particular document created by a user will hold many duplications and therefore be subject to the coupling and ripple effects as described above.

Recently some software solutions have provided users with some ways of keeping track of different versions of documents in a more detailed way. For instance, most present-day word processors have a "track changes" functionality in which changes to sentences can be proposed by users and inspected, accepted or rejected by others. These are however mostly limited to a single concurrent editor of the document as of today. Furthermore tracking of changes and variants in these software packages is limited to the scope of the overall document and tracking a long history of document changes is cumbersome.

For these reasons, several more advanced version control systems (e.g., git or subversion) have been introduced over the past decade. These systems allow users to keep track of document changes made by multiple concurrent users over time. The users can commit to versions and merge their versions with changes in versions of other users in an often automatic and much more user-friendly way. These systems also support document variability by allowing different branches to be created of documents, cherrypicking, etc. Although these version control systems provide well desired functionality, they are mostly incompatible with the most frequently used and recently released word processors or presentation applications. This is because in order to work at their fullest potential (i.e., allowing merging, track changes within individual files, etcetera) these systems should work with textual files (such as .txt, .tex, .md, or .java). More popular file extensions (such as .doc, .docx., .ppt, etcetera) are however binary and do not support universal and granular version control by external systems. Because they work on textual instead of binary files, the advanced version control systems (e.g., git or subversion) work at the level of the individual lines within files. As such, documents are considered at a very fine-grained modular level (i.e., individual lines). However, this level does not correspond to a meaningful, anthropomorphic modular level such as the ones we mentioned in the paragraphs above (e.g., paragraphs, part descriptions, clauses, etcetera).

Contrary to these approaches, the modular view of documents we propose in this paper allows for version control on a meaningful level: based on changes in the text modules. As such a module contains information on a specific aspect (i.e., a content cross-cutting concern), changes to this information should be contained within the module. Therefore the decomposition of information into text modules can also be guided by separating all change drivers (i.e., information that can change independently) into separate text modules.

This approach to version control detaches the management of versions from content-agnostic levels (such as the document or document lines) and performs it on a more useful level based on the actual content of textual components. As such, the version of a document becomes based on the total composition of the text module versions it contains. This also entails the simple chronological and sequential way of versions control is a thing of the past. Instead, we can manage the different versions of document that exist in two dimensions: time and content.

First, managing the successive versions of documents is lowered to the level of text modules instead of an entire document. The current version of this document is simply the aggregation of all versions numbers of the text modules it contains. Take for example the scenario in which two text modules where updated and a new version of a document was generated. However, one of these module updates was premature and needed to be rolled back. In a chronological and sequential version control system, the new correct document would simply get a higher version number assigned to it. As such, a lot of information on these changes would get lost when it is not captured in semi-structured or unstructured version information. In the modular approach the version numbers are managed on a text module level, which clearly shows what components of the document were updated and rolled back in the successive document versions.

A second dimension of variation in modular documentation is introduced by the free composition of text modules into new types of documents. This means variations of documents with minor differences can be easily generated and managed. An information system would need to be constructed that keeps track off all these versions of documents and updates those documents that need updating when a new version of a text module is introduced. Taking into account the Version Transparency principle discussed earlier, some documents should however be able remain unchanged because they might need to contain the older version of the text module. This approach of version control requires several types of functionality:

– Maintaining a version history of all text modules
– Storing all actual past and current versions of text modules
– Keeping track of which documents the text modules are used in
– Keeping track of which text modules the documents are made up off
– Assimilating text modules changes to all documents that need to be updated

It is clear from this list that such a version control system is more complex than a simple linear and sequential version control system.

5.3   Relative Sectioning

Another important aspect of allowing text modules to be re-usable is to implement relative embedding. This is the possibility to embed a text module in the hierarchical structure of overarching documents in a "relative" way (e.g., defining that a particular piece of text should be considered as contained within a section situated one level deeper as its surrounding text: this could be subsection or subsubsection in LaTeX, a H1 or H2 section in an Markdown document, etcetera). As such, text modules can be used in all kinds of documents and on all kind of document levels without this requiring any change in the text module.

The LaTeX document preparation system for example allows the hierarchical inclusion of sub-files (i.e., text modules) and allows the layout cross-cutting concern to be handled in a separate layout file. LaTeX however does not provide a system for relative sectioning out of the box. The hierarchical structure of

```
\input{input/"Content description"}
\leveldown
\input{input/"Business Economics"}
\leveldown
\input{input/"Business Economics_Accountancy_content "}
\input{input/"Business Economics_European and International Law_content "}
\input{input/"Business Economics_Finance_content "}
\input{input/"Business Economics_Marketing_content "}
\input{input/"Business Economics_Strategy and Organisation_content "}
\input{input/"Business Economics_Transport and Logistics_content "}
\levelup
\levelup
\input{input/"Assignments"}
\leveldown
\input{input/"Business Economics"}
\leveldown
\input{input/"Business Economics_Accountancy_assignments"}
\input{input/"Business Economics_European and International Law_assignments "}
\input{input/"Business Economics_Finance_assignments "}
\input{input/"Business Economics_Marketing_assignments "}
\input{input/"Business Economics_Strategy and Organisation_assignments "}
\input{input/"Business Economics_Transport and Logistics_assignments "}
\levelup
\levelup
...
```

Fig. 1. Example of a LaTeX Structure File supporting relative sectioning

sections (i.e., whether something is a chapter, section, subsection, subsubsection, etcetera) should be hard coded within .tex files and therefore limits the potential for text modules to be freely combined into final documents which might use the same text excerpts at different levels within their own document hierarchy. To overcome this problem, a LaTeX style file can be used that provides the functionality of relative sectioning [6]. This allows the generation of a LaTeX structure file, of which the first part is shown in Fig. 1. In this file, text modules are imported via the \input{} command. The names included in this command are the files that should be part of the generated document. More importantly, the \leveldown and \levelup commands can be automatically added whenever the next text module of the document should be added on a lower or higher level. As such, the basic text modules exist of solely a title (included within the \dynsection{} that is provided by the custom style file) and the content of the module.

5.4 Dynamic Cross-Referencing

The re-use of text modules also requires an alternative to static cross-referencing. Because text modules can be inserted in a document on different locations or levels, they do not have fixed section numbers across documents. For this reason, modular documents need dynamic referencing to refer to specific text modules. This can be achieved by inserting a static reference point in every text module. In LaTeX, this can for example be achieved by using the \label{} command, in which a label can be defined for a section. This section can than be referred to by the `\ref{}` command. These commands can be used to reference to text modules, figures, tables, equations, etc.

Using this cross-referencing system however requires additional checks to be incorporated. Because of the combination potential of text modules, some references might refer to components not included in the document. This would cause undefined references in the document. To prevent this, the referencing system should include some mechanism that checks whether the referenced component is included in the document. If so, the reference can be included. If the reference point cannot be found, the sentence referencing to it should not be included in the document.

6   Modular and Evolvable Documents in Practice

6.1   Decomposing Documents into Text Modules

Based on the aspects of modular and evolvable design of documents, a prototype was developed to show the practical feasibility of a system that supports this document management approach. This has been discussed in more detail in previous research [9]. The prototype manages documents concerning study programs at a university. Being the underlying artifacts, the study programs' architecture was first modularized and made evolvable. After this redesign was finished, existing content describing the courses was looked at and modularized to allow the generation of different kind of documents. Information on the study programs was derived from the course descriptions available on the faculty's website. From these descriptions, text modules with similar content were identified based on the change driver guideline. This lead to the definition of 10 types of text modules, based on 10 content cross-cutting concerns that were identified to be present in courses and study programs. These concerns include for example a short content description, internationalization, lended learning, assignments, etcetera. Combined, these 10 types of text modules allow for a complete representation of a course. As we defined learning-teaching tracks and study programs to be aggregations of courses according to modularity reasoning, the text modules can be used to represent these parent artifacts as well. Taking into account the total number of 258 courses and 10 content cross-cutting concerns, the modularization of the course descriptions resulted in a total of $258 * 10 = 2,580$ text modules. These text modules can be used to represent all aspects of the courses, learning-teaching tacks and study programs of the faculty.

Although this large amount of text modules seems inefficient to manage and maintain, this strict decomposition actually simplifies several aspects of document management. First, this imposed separation of concerns established a recurring structure across all course descriptions. This gives professors (who are responsible for the content of the text modules) a template to hold on to in describing their courses. Furthermore it is easier to retrieve information, as it is separated in meaningful text modules that are named after the concern they include. This also allows for easier maintenance of the information. However, the most important advantage of the decomposition is the vast amount of possibilities combinations of modules that can be generated into documents. The information included in the decomposed text modules allows for the generation of a vast variety of documents with different purposes. This system for example allows one to generate documents containing the assignments of all courses in a study program. But the system can also generate a document listing all courses or learning-teaching tracks in which some sort of internationalization is present. But the true value of the systems becomes even clearer if for example students were to be added. This would allow the system to generate for example a document detailing all sustainability or social impact aspects a student has encountered in his study program. Or how much hands-on experience he has gained due to assignments or case studies. Such a system would allow to draw up student-specific diploma's with one click of a button. In general, the decomposition thus allows for a versatile use of document modules and allows the definition of new types of documents with new purposes.

6.2   Document Versatility, Variability, and Evolvability

As mentioned, the modular structure of documents provides an amount of *document versatility*. Let's explain this in numbers. Take for example a faculty that offers 12 study programs (5 Bachelor and 7 Master programs). For each study program, we would like to be able to generate a document consisting of two or three document levels. If we abstract from the course level to simplify the calculations, there are 3 possible selections for the first document level (i.e., cross-cutting concerns, learning-teaching tracks and sub-tracks). Therefore there are only two selections left for the second level (the two remaining ones), and two possible selection for the final level (i.e., either choosing the remaining selection or not including a third level). This totals up to 12 possible selections for the document levels. If we consider either including or not including the 10 cross-cutting concerns, the amount of combinations adds up to $2^{10} = 1024$ possibilities. Multiplying the 12 study programs, 12 possible document level selections and 1024 possible combinations of 10 cross-cutting concerns inclusions gives us a total of possible document combinations that can be generated based on the 2,580 defined text modules:

$$12 * 12 * 1024 = 147,456 \text{ possible documents} \tag{1}$$

However, if for example 3,000 students were to be included in the system, the document versatility would increase exponentially. Let's assume there are 1,000 unique versions of study programs of these 3,000 students (which is actually a cautious estimate considering the amount of courses students can choose in some study programs). Substituting the 12 study programs by 1,000 study program versions in the previous multiplication results in the total amount of possible document combinations:

$$1000 * 12 * 1024 = 12,288,000 \text{ possible documents} \tag{2}$$

This example clearly shows the combination potential of decomposing course descriptions into fine-grained text modules.

As mentioned earlier, the decomposition in text modules also allows more fine-grained version control to manage the *variability* in all types of documents that can be generated. Changes can be tracked more specifically when version control is managed on a text module level. A text module can be archived based on their moment(s) of change, allowing the generation of documents according to specific time specifications. This can prove to be very useful. Imagine for example the re-generation of a student diploma after it has been lost. The graduation might have been a few years ago, so courses and study programs will have changed. Yet it is important for a university to be able to generate the diploma with the correct descriptions of the version of the courses the student took. This is possible with a module-based version control systems that takes into account both versions in time and concurrent variations. This example shows the importance of tracking changes on a fine-grained modular level.

Furthermore, the use of modular text modules shows the importance of eliminating combinatorial effects to achieve *evolvability*. Any change in the description needs to be made in only one of the 2,580 files/text modules. By creating a script that regenerates all documents in which this module is included, this change is easily applied to all documents it is included in. As such, combinatorial effects are avoided and evolvable documents can be generated.

7   Conclusion

In this paper, we presented an alternative approach to the use of static and monolithic documents. Compared to traditional document and version management systems, this new approach allows for more versatile document management. In this paper we have shown how documents can be viewed as modular systems and decomposed into text modules. There are several advantages to this approach. First, modular documents are easier to maintain. The text modules they are composed of show a clear structure and clearly indicate the single point of storage of specific information. Another important advantage is the vast versatility the use of text modules creates. New types of documents can be composed simply by combining text modules in new ways. As such, new types of documents can be created with goals and purposes the user might now even be able to imagine right now. This is shown in the paper by calculating the number of possible document combinations for study program documents. And finally, the systematic decomposition of documents into text modules allows the creation of evolvable documents. By adhering to the Separation of Concerns and Version Transparency principles, combinatorial effects can be eliminated and evolvable documents can be generated.

Another contribution of this paper is the discussion on dynamic cross-referencing, relative sectioning and version control. The paper shows the importance of these aspects in document management and in obtaining evolvable documents.

In future research, additional cases will be studied to corroborate the theoretical and practical findings of the study program case mentioned in this paper. Additional research will also be done to potentially expand the list of identified cross-cutting concerns in documents.

References

1. Baldwin, C.Y., Clark, K.B.: Design Rules: The Power of Modularity, vol. 1. MIT Press, Cambridge (1999)
2. Campagnolo, D., Camuffo, A.: The concept of modularity in management studies: a literature review. Int. J. Manage. Rev. **12**(3), 259–283 (2010)
3. Huysmans, P.: On the feasibility of normalized enterprises: applying normalized systems theory to the high-level design of enterprises. Ph.D. thesis, University of Antwerp (2011)
4. Gruber, J.: Markdown (2004). https://daringfireball.net/projects/markdown/
5. MacFarlane, J.: Pandoc Universal Document Converter (2017). http://pandoc.org/
6. Leichsenring, C.: Relsec style file (2013). https://github.com/mudd1/relsec/blob/master/relsec.sty
7. Mannaert, H., Verelst, J., De Bruyn, P.: Normalized Systems Theory: From Foundations for Evolvable Software Toward a General Theory for Evolvable Design. Koppa (2016)
8. Mannaert, H., Verelst, J., Ven, K.: Towards evolvable software architectures based on systems theoretic stability. Softw. Pract. Experience **42**(1), 89–116 (2012)
9. Oorts, G., Mannaert, H., Franquet, I.: Toward evolvable document management for study programs based on modular aggregation patterns. In: PATTERNS 2017: The Ninth International Conferences on Pervasive Patterns and Applications, p. 6 (2017)
10. Porter, M.E.: Strategy and the internet. Harvard Bus. Rev. **79**(3), 62–78, 164 (2001)
11. Van Nuffel, D.: Towards designing modular and evolvable business processes. Ph.D. thesis, University of Antwerp (2011)
12. Vanhoof, E., Huysmans, P., Aerts, W., Verelst, J.: Evaluating accounting information systems that support multiple GAAP reporting using normalized systems theory. In: Aveiro, D., Tribolet, J., Gouveia, D. (eds.) EEWC 2014. LNBIP, vol. 174, pp. 76–90. Springer, Cham (2014). doi:10.1007/978-3-319-06505-2_6

Application of Enterprise Engineering to Lean Process Management: An Explorative Case Study

Marjolein Deryck[1(✉)] and Philip Huysmans[2(✉)]

[1] Faculty of Applied Economics, University of Antwerp, Antwerp, Belgium
marjolein.deryck@uantwerpen.be
[2] Antwerp Management School, Antwerp, Belgium
philip.huysmans@ams.ac.be

Abstract. Since the publication of 'The machine that changed the world', lean has spread in academic and management literature. In this case we study the lean transition of one process in a Belgian financial institution by means of Enterprise Engineering (EE) concepts and methods. Enterprise Ontology (EO) and Normalized Systems (NS) are used. Additionally, Business Process Modeling Notation (BPMN) is used to describe process implementations in more detail. In the case the role of each of these methods is discussed, as well as their weaknesses and advantages. However, it does not propose an explicit integration of different theories. Based on our results and in line with calls from previous research we highlight the contributions of EE tools for lean implementation.

Keywords: Enterprise Engineering · Ontology · Normalized Systems

1 Introduction

A bulk of management and academic literature is available on the advantages of introducing lean principles in the organization (see e.g. [8]). The huge reported cost reductions reported by some manufacturers spurred its diffusion to virtually every sector. The effective realization of benefits however don't seem to fulfill the high expectations [3,7]. The absence of a solid implementation methodology might partially explain this disappointment [5]. This is especially true for the application of lean in services sectors, for which the development of frameworks only started recently. Moreover, it is recognized that the application of lean in these sectors requires different tools, because of the intangible nature of services, and their intensive use of Information Systems (IS) [4].

EE may provide an answer. The discipline views organizations in analogy with other engineering disciplines as a system that needs to be designed purposefully [1]. Hence it develops tools and practices to systematically (re-)design organizations. The case study research presented in this paper focuses on the lean transformation of one of the operational processes started in a Belgian financial institution. The difficulties that the enterprise encountered during the first lean

© Springer International Publishing AG 2017
D. Aveiro et al. (Eds.): EEWC 2017, LNBIP 284, pp. 141–148, 2017.
DOI: 10.1007/978-3-319-57955-9_11

phases were considered opportunities to connect with the contributions of other methods. This resulted in the description of how different methods can be used to achieve the predetermined purpose of a business process optimization project. A related set of ontological models provides the starting point to differentiate between value-adding and other activities, to map information streams and to develop the data model of an IS. Our research also included the construction of a prototype based on the data model. It can be used to visualize and discuss the work in progress with end users and programmers. It also demonstrates how NS are built to cope with changes and complexity.

This paper starts with the literature background of lean and EE. Section 3 discusses the case study: methodology, the background, and the contributions of EO, BPMN and NS. The implications of these results are discussed in Sect. 4. Finally the conclusion focuses on contributions and limitations of this research.

2   Related Work

2.1   Lean and Challenges

Lean appeared in academic and management literature at the end of last century. It generalized the production principles from Japanese car manufacturers that succeeded to produce more and better cars with less resources [8]. Central in lean is the idea of 'waste' i.e., all activities that don't add value to the customer.

At the end of last century some of the research focus shifted towards the application of lean in service sectors. These sectors account for the largest part of the GDP in most developed countries, but their efficiency lags far behind that of physical manufacturing processes [9]. Nevertheless, a literature survey reveals that many challenges remain when implementing lean in service sectors.

E.g., a generally accepted framework explaining **how lean can be implemented seems to be lacking** (see e.g., [5]). Some frameworks prove useful in a single case but don't succeed in conveying the advantage to others.

This seems to be even more true in a service context [9]. A recent framework is 'Lean first, then automate', by Bartolotti and Romano [4]. Preliminary to the development of the framework their research focused on the relation between automation and lean. Traditionally lean advises to avoid the automation of processes, as this is a source of rigidity. However, in service processes often the efficiency and quality increase when performed automatically. Therefore automation in a pure service context is desirable, but since the **automation of inefficient processes is counterproductive**, processes first need to be streamlined [4].

To implement frameworks they need to be supplemented with practical tools. Previous research indicates that **the nature of these may differ from that of manufacturing tools**. E.g. the determination of waste is acknowledged to be particularly complex due to the intangible nature of services [2].

One of the techniques usually associated with lean is brainstorming. By uniting members from different teams and having them put forward possible ideas, a solid solution will be attained. However, in practice the quality of the solution **depends heavily on the knowledge of brainstorming participants**, personality traits and organizational influences.

2.2   Proposal: Purposeful Design with Enterprise Engineering

One possible remediation for these challenges is to call for a purposeful design of the enterprise as put forth by the research area of EE. EE provides a set of tools and theories to this end. It is not the intention of this paper to propose an explicit integration of different theories. Nevertheless, the engineering perspective which is the basis for all EE theories provides a common ground to contribute to the lean methodology in complementary parts of a single case.

Enterprise Ontology provides a structured way to create models by the selection of only the essential (ontological) transactions. The theoretical underpinning is rooted in Habermas' communicative action theory. It builds on the similarity between human communication and enterprises as social systems with multiple organizational layers. This is discussed extensively in the work of Dietz [6]. Starting from EO models, the aim is to deduct an optimal implementation from it.

The two main constructs used from EO are transactions and actors. Ontological transactions are the basic building blocks of EO models. Abstraction is made from the way these transactions are implemented in practice, which greatly reduced the complexity associated with it. As a result, EO models are suitable tools to enhance comprehension of clients and participants with regards to the end-to-end process.

An actor role in EO is a theoretical construct that combines competence, responsibility, and authority [6]. Mapping the ontological actor roles with the implemented organization functions can provide insight in difficulties linked to the information streams and responsibilities.

BPMN stands for Business Process Model and Notation. It is a widely used standard to translate business processes in a visual representation. BPMN is not a part of EE, but was used in this project to describe process implementations in more detail, and to be able to integrate with the current modeling efforts of the team. It supplements EO models with other aspects of organizational actions situated at the infological level. Because of the intuitive way in which BPMN models can be read and their wide-shared business use, it was preferred over other process models.

Normalized Systems prescribes a set of design principles to achieve evolvability at the implementation level, which is operationalized as the absence of combinatorial effects (CE). While the theory is applicable to different domains, its application is most advanced in the software domain. The NS design principles are adhered to by proposing software elements, which each align with these four theorems. These elements provide building blocks which enable a fast expansion of flexible and scalable IS. These expanded systems can be used as analysis prototypes, which allow a very iterative way of fine-tuning business requirements of an IS. Moreover, the four principles can be used to identify and explain the causes of agility problems linked to the current way of working.

3   Demonstration: Case Study

3.1   Case Study Methodology

This research is conceived as a single descriptive case study of the lean transformation of a cross-departmental processes. It illustrates the challenges and pitfalls that initiators need to overcome if they aspire to reap all of lean's benefits. Information was gathered through analysis of internal procedures and documents, and supplemented with interviews. Minutes of every interview were drafted and hence reviewed and (where needed) supplemented by meeting participants. Drafted models, preliminary insights, and conclusions were discussed with the lean tracker to ensure alignment with business insights. All of this information was stored in a case study database. For each of the methodologies used, it is indicated in the relevant section how the researchers proceeded. The research started with the selection and demarcation of th process. It continued along the analysis, brainstorming phase and the selection of optimization tracks. Due to time constraints the effective implementation phase was not covered.

3.2   Case Study Background

The case study company is a financial institution in Belgium. It decided to initiate a lean transformation in a selected scope of departments. They were encouraged to select one process that cuts across different departments and in which lean optimizations were to be expected. Lean was selected as method because of the central position of the customer needs and to promote a general "lean attitude" among employees.

The case study department is the back office financial markets. A selection of tasks performed by this department include amongst other the validation of deals concluded by the trading room, execution of payments, follow-up on collateral, and financial and liquidity reporting.

The selected process has the purpose to conclude master agreements (MA) that facilitate the conclusion of financial market transactions. The MA describes the general conditions, restrictions and practices applicable to the settlement of transactions. This way it suffices to confirm the specific details when an actual transaction between the two parties is concluded (typically resulting in a three-pages deal confirmation), given the fact that all other processing details are agreed upon in the MA.

The targeted areas of improvement are the communication with customers and internal communication between departments.

3.3   Results

Enterprise Ontology *Procedure.* To ensure the collection of all relevant information and validity of the conducted analysis we proceeded through five steps of procedure analysis, interviews, ontological analysis of the obtained information, grouping of transactions and validation. It is worthwhile to elaborate on the validation step. In an earlier meeting the concepts of ontological transactions were briefly (+− 30 min) explained to the business owners. Before the presentation of the drafted models the main principles were repeated. The validation of the construction model and the explanation of some pieces of information missing in the detailed process models of the transactions took 90 min. The validation of the MA process model took less than 15 min, including the repetition of modeling principles. As a result of this step minor corrections and replenishment details were made.

Contributions. The ontological models consist of departmental and process models. The departmental models have the department as its universe of discussion. They contain the transactions performed or initiated by the department and the resulting detailed models. The process-induced models have the MA process as its universe of discussion. The connection between these two kind of models is a shared transaction in which the back office acts as an actor.

The departmental models allowed for **insights on the organizational role** of the department. By mapping the ontological model with the organization chart it became apparent that given its size the number of transactions is very small. Further analysis reveals that an even smaller number of teams execute these transactions. Moreover, the transactions are linked to the mandatory distinction between the trading room and back office. This may hamper the generalization to other back-offices that may not have similar legal requirements.

The process models provide more **insight on the actors in the process** and the role of the back office in it. The set of ontological process models (process structure diagram, action model, information use table and state model) make up a solid base for the design of the related IS. In the elaboration of the detailed process models an additional mapping of actor roles on organization roles has been performed for analysis purpose. This leads to the conclusion that in the current organization a fragmented communication to the customer is almost inevitable, as the actor role responsible for it is taken up by multiple teams in the process chain.

Also a **result-structure chart** has been created. This model provides a very high-level overview of the process. Therefore it is an ideal way to provide insight in the broader purpose of the requested actions. Also, this model highlights in a very condensed format the activities that need to be performed successfully to conclude the contract with the customer. In lean-terminology this equals the value-adding activities of the process. Therefore it is very suitable to extend the regular lean tools in service processes.

BPMN *Procedure.* The drawing of the model based on an operational procedure proved difficult due to the lack of process structure contained in it. The document available for the creation of the BPMN model was an eight pages long operational procedure that described in a structured way the phases in the process, the executor, the input, the activities and the output. The high level process steps mentioned in the procedure did not contain enough information to use as a guideline. The detailed operational information on the other hand did not discern important activities from subactivities. These difficulties could be overcome by enriching the drafted EO process models with implementation details on info- and datalogical level to create the BPMN model.

Contributions. The BPMN model revealed that the procedure signed off between the different departments mainly focuses on the successful processing of the MA request. The tasks to perform with non-standard cases are not described, while probably this type of files make lead times surge. Hence the explicit inclusion of deviant situations could diminish average lead times. Moreover, it became apparent that some information flows were not well documented, this in spite of its ambition to include all information flows in the procedure. Therefore BPMN is a useful tool to check the completeness of step-by-step procedures. Additionally, using actor swimlanes might stimulate cross-departmental process-thinking. The model might also be useful to identify suitable process owners and relevant process performance indicators.

Normalized Systems *Procedure.* One of the optimization tracks concerns the automation of the MA input information file. First the motivation to automate the way of working were analyzed using NS theorems, then an evolvable prototype of an IS was created. The information file contains multiple worksheet and some very detailed information on the client and his request. The big blocks of the data model and the links between them were identified in the state model. For the MA process these blocks are: client, masteragreement, transaction, credit line, and guarantor. The information in the sheet contains more detailed attributes. After assigning the attributes to one of the blocks, the obtained data-objects were evaluated against the four NS theorems, thereby splitting the client-object in multiple data-objects. This procedure was followed for each of the five original blocks. The result is a NS-compliant data model that can be used as input for the creation of the application structure using NS expanders.

After the data-structure has been built, a workflow representing the process are identified and inserted into the data structure. Without further development the application can be used as such, with employees performing tasks manually and adapting the status accordingly. Company-specific customization code can be inserted to automate some of these tasks. This code can't be guaranteed to be free from CE, but impacts will be limited to the structure in which it is encapsulated.

Contributions. The evaluation of the sheet against the four NS-theorems explains and underpins the factors that inhibit evolvability. A clear view on this might prove helpful to substantiate project benefits. By applying these same criteria to the design of the data model it forms the basis of the stable application structure. The way to obtain this is straightforward and described above. In the situation where a new application needs to be created, not all requirements may be available beforehand. In that case a more basic prototype can easily be developed based on the building blocks identified in the state model. This in turn can be used to solicit feedback from end users and reveal additional requirements. As the prototype is evolvable, new requirements can easily be included in it. Therefore it is an easy tool for an analyst to create an evolvable application structure, to collaborate on short development cycles with end-users without a lot of formalization. This complies well with the aim of lean projects to eliminate waste, as the laborious requirements analysis phase and its translation in documents is replaced with quick and tangible results.

4 Discussion

This paper starts from the lean transformation of a process in a financial institution. By using different widespread methodologies on the same process the contributions of each method become apparent. For this discussion, we refer to the challenges for lean identified in Sect. 2.

A first challenge referred to the lack of a generally accepted framework. The absence of complicated models and the use of brainstorming techniques are seen as facilitators to implement lean. However, in practice the absence of a standardized implementation and tools induces difficulties with regards to the selection and demarcation of the process to tackle, insight in the construction of the process and translation to technical requirements for supporting applications. EE consists of a set of interrelated theories, rooted in a common engineering perspective. This can be illustrated by the way the different EO aspect models are used as a basis for modeling the BPMN processes and NS data models. This reuse of knowledge captured in a certain model also demonstrates how the challenge regarding the heavy dependence on brainstorming sessions can be addressed. Moreover, it was illustrated how the consistent usage of and reference to the transaction model allow a demarcation of the brainstorming session, which allows for more focused suggestions.

The explicit usage of EO models as a starting point in our approach highlights the concern to address the challenge regarding the need for different lean tools in a service context. EO considers an organization as a social system, which provides an adequate answer to the complexities in service organizations. The demarcation of responsibilities of the department as discussed in Sect. 3.3 illustrates this point. This illustration shows that the creation of ontological models does not hamper the dynamic and participation-stimulating nature of lean. On the contrary, it may focus group discussions and drawing attention to possible conflicts of interest. Because of the specific nature of ontological transactions they can also be considered as the value-adding activities in a process. Hence the ontological process model is particularly suitable for use in the initial stage of a lean transformation that embarks with the identification of waste.

Finally, the inclusion of the NS prototype discussed in Sect. 3.3 enables flexibility and elicits experimentation for the proposed solutions. E.g., it was mentioned how the workflow constructs of NS allow interchanging manual or automated task implementations. Consequently, our approach enables a focus on the overall process structure first, and postponing discussions regarding automation. Additionally, the focus of NS on evolvability forces the artefacts of a lean project to be prepared for future alignment with new customer preferences.

5 Conclusion and Future Research

The case study presented in this paper contributes to the body of knowledge in three ways. First, it illustrates the practical application of EE concepts in a real life case. Second, it contributes to the expansion of lean that was called for by other authors [8]. It does this in such way that it copes with the challenges associated with the use of lean in service sectors. In our study we show how EO models can be used to understand the organizational structure, frame the initial discussion and guide the subsequent design. On the other hand NS prototyping associates well with lean in the implementation phase, as it allows quick and easy modeling in short runs and without the waste typically associated with requirements definition. Third, this paper extends the evidence that it is possible to design processes and IS based on engineering principles. This implies a more deterministic approach in which factors like experience of the modeler or the coincidence of a successful brainstorming idea are less important.

References

1. Albani, A., Rabera, D., Wintera, R.: A conceptual framework for analysing enterprise engineering methodologies. Enterp. Model. Inf. Syst. Architect. **11**, 1 (2016)
2. Andres-Lopez, E., Gonzalez-Requena, I., Sanz-Lobera, A.: Lean service: reassessment of lean manufacturing for service activities. Procedia Eng. **132**, 23–30 (2015)
3. Bhasin, S.: Prominent obstacles to lean. Int. J. Prod. Perform. Manage. **61**(4), 403–425 (2012)
4. Bortolotti, T., Romano, P.: Lean first, then automate: a framework for process improvement in pure service companies. a case study. Prod. Planning Control **23**(7), 513–522 (2012)
5. Chay, T., Xu, Y., Tiwari, A., Chay, F.: Towards lean transformation: the analysis of lean implementation frameworks. J. Manuf. Technol. Manage. **26**(7), 1031–1052 (2015)
6. Dietz, J.: Enterprise Ontology: Theory and Methodology. Springer, Heidelberg (2006)
7. Pepper, M., Spedding, T.: The evolution of lean Six Sigma. Int. J. Qual. Reliab. Manage. **27**(2), 138–155 (2010)
8. Samuel, D., Found, P., Williams, S.J.: How did the publication of the book the machine that changed the world change management thinking? Exploring 25 years of lean literature. Int. J. Oper. Prod. Manage. **35**(10), 1386–1407 (2015)
9. Surez-Barraza, M.F., Smith, T., Dahlgaard-Park, S.M.: Lean service: a literature analysis and classification. Total Qual. Manage. Bus. Excellence **23**(3–4), 359–380 (2012)

Ontologies

The REA Model Expressed in a Generic DEMO Model for Co-creation and Co-production

Frantisek Hunka[1(✉)] and Steven J.H. van Kervel[2]

[1] University of Ostrava, Ostrava, Czech Republic
Frantisek.hunka@osu.cz
[2] Formetis Consultants BV, Boxtel, The Netherlands
info@formetis.nl

Abstract. The REA ontology is a domain ontology that aims to support accounting information systems that must provide a truthful and appropriate – GAAP compliant - descriptive perspective of an enterprise in operation. While the application of a domain ontology provides strong benefits, the current representation of the REA model does not provide the desired results; an appropriate working accounting system. One of the root causes of this problem is the lack of a proper formal representation of the REA model. In this paper the DEMO methodology is applied to provide a generic domain and application-independent DEMO model (the CC-CP model) for co-creation and co-production in any industrial production chain. This model appears to be also appropriate to capture any interaction between an enterprise and any external parties, stakeholders, customers, suppliers, personnel etc., and support accounting systems. This approach offers several new advantages, notably: (i) prescriptive workflow-like operation of the enterprise with full transaction driven execution; (ii) process-mining (-like) analysis of daily operation; (iii) ontological completeness of factual knowledge as required not only for accounting systems but also for other descriptive information systems and (iv) completeness of implementation for any kind of business interactions between enterprises.

Keywords: REA model · DEMO enterprise ontology (DEO) · DEMO methodology · Co-creation and Co-production

1 Introduction

REA is considered a strong potential improvement of foundations for accounting systems. It aims at providing a domain ontology which is a necessary condition of any system that provides some perspective of "phenomena in the real world". However, so far any attempts to apply REA for accounting systems using the current representation [8, 9] have encountered serious problems that are difficult to identify and even more difficult to fix [11, 12]. Its present formal representations appear not appropriate enough [12]. It is argued and observed that the root of these problems is obviously caused by: (i) a lack of good ontological foundations; (ii) lack of good formal methods and (iii) lack of an appropriate formal language to represent the model.

© Springer International Publishing AG 2017
D. Aveiro et al. (Eds.): EEWC 2017, LNBIP 284, pp. 151–165, 2017.
DOI: 10.1007/978-3-319-57955-9_12

For the study and understanding of phenomena observed in reality, ontologies are designed and used to understand and reason about some specific domain of reality (reality "precedes" ontology). Section 5 specifies REA model-driven GAAP compliant systems and the core theories. An ontology is defined as a "formal, explicit specification of a shared conceptualization of reality" (Gruber 1993). The conceptualization is composed of concepts of objects, their attributes and their relations. The conceptualization is shared by knowledgeable human stakeholders, meaning in the first place that all stakeholders are assumed to have an identical understanding of each object or concept, each attribute and each relation. Informally, an ontology has two faces; one face to the phenomena in the real world, and the other face a formal explicit representation of concepts.

The REA ontology must capture all phenomena in the real world that are required by some accounting system to provide a GAAP compliant accounting perspective and representation of that enterprise. GAAP requirements define which phenomena must be captured with a completeness criterion. If not all phenomena are captured or if the phenomena are not captured in a truthful way, then any accounting system will provide a wrong perspective, a wrong profit -loss statement, general ledger etc.

In this paper a representation of the REA model expressed in a DEMO model for co-creation and co-production (CC-CP) [12] is assessed. The CC-CP model is generic, application and industry independent and captures all relations between suppliers, stakeholders and individual workers for our enterprise of interest. This claim demands future empirical support.

In Sect. 2 the ontological foundations of REA and DEMO are assessed. In Sect. 2.1 the strength and weaknesses of the REA ontology are described and why another representation of the REA model is needed. Notably the problematic notion of "value" has been addressed and solved [3]. In Sect. 3 the CC-CP model is presented and assessed. Section 4 describes the benefits of the CC-CP model representation of the REA model and several quality and completeness criteria.

Future research, our long term strategic vision and objectives are described in Sect. 5. If we have a proper formal representation of the REA model that is provided by the DEMO CC-CP model, then that is a promising foundation for future model driven GAAP compliant accounting systems, with strong benefits.

2   Ontological Foundations of REA and DEMO

The ontological foundations, the strength and weaknesses of REA are assessed. Its apparent weaknesses are notably lack of formal methods, lack of empirical theories and ontological flaws and incompleteness [14]. The strengths of the DEMO Enterprise Ontology (DEO) are described; notably strong empirical foundations, formal methods, design science and general systems theory. This approach is to express the REA ontology in concepts and relations provided by the DEO in such a way that strength and value of REA is kept and the apparent weaknesses are mitigated.

2.1   REA Ontology

The REA ontology originates from accountancy systems and provides a domain specific platform for value modeling business processes, see [10]. The principal economic concepts are economic resources, economic events and economic agents. Economic resources are things of economic value that have utility for economic agents and for this reason they used to be planned, monitored, and controlled. Economic events are activities within an enterprise that represent either an increment or a decrement in the value of economic resources. Economic agents are individuals or organizations that participate in the control and execution of economic events.

The other fundamental REA concepts, their relationships, constraints and rules for constructing application models are illustrated in Fig. 1. Apart from the above mentioned concepts, Fig. 1 also contains commitment and contract concepts and corresponding relationships.

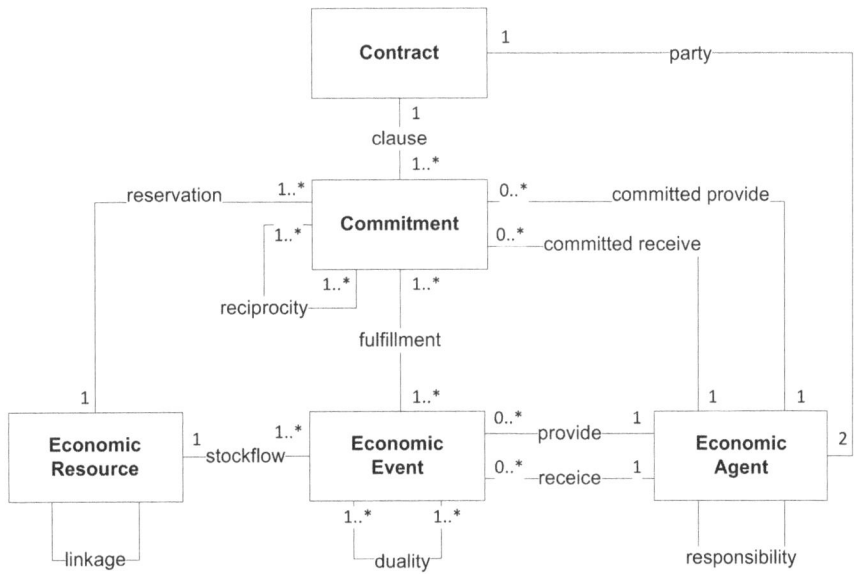

Fig. 1.   REA metamodel level. Adopted from [9]

The main benefit of the REA approach is that all accounting artifacts such as debit, credit, journals, ledgers, receivables, and account balances are derived from the data describing exchange and conversion REA processes. It means that all accounting artifacts are always consistent, because they are derived from the same data; for example, data describing a sale event is used in warehouse management, payroll, distribution, finance and other application areas, without transformation or adjustment [9].

REA anomalies have its origin in absence, to some degree, of rigorous theoretical and philosophic foundations, mainly since these seem to be lacking. One of these lacking capabilities is that the REA model itself does not have specific states from which a state machine can be derived. Observation shows that in REA between two entities there are different discrete, disjoint, states of any economic transaction. Instead, only the resource states are identified and frequently used as the states of the state machine, which is incorrect. As a result of this, the REA model cannot provide decline and reject transaction steps, as well as revoking operations, things one observes happening in any economic transaction. In addition, the REA model is predominantly design to capture events that refer to the resource value or resource feature. The other events such as business events or information events are difficult to capture and further processed. Consequently, REA has not properly defined so called information or knowledge entities such as contract or schedule because these entities are not resources. REA contains a contract concept which represents a contract that came into effect. There is no concept in REA for unsigned (not coming into effect) contract. The REA modeling approach aims at descriptive information systems that are based on exchange, consumption, usage and production of economic resources.

To summarize the main observed limitations and flaws of REA in its current representation.

1. Restriction to capture only production facts that in addition refers only to the exchange of property rights to resources or to transformation of a set of resources into another set of resources.
2. Lack of ontological completeness of a *transaction* for example: sending a production order, receiving an invoice etc. These are all important facts for accounting systems. Yet not provided by the current representation [9].
3. Not capturing *transaction* events which imply that there is no truthful state machine based on transaction states and state transitions. This also includes missing transaction steps such as decline or reject, as well as cancellation patterns.
4. Conceptual mismatch – not capturing the real phenomena in the world. It is manifested in explicit distinction between past and current events and events which are performed in future and in impossibility to express the change of state in which a contract or a schedule comes into effect.
5. Restricted ability to express explicitly business rules. The type level mechanism applied in REA enables only to impose business rules on the instances which are in compliance with the given type.
6. No prescriptive capabilities – no certainty that "in real life" things go as defined.
7. REA and accounting apply the notion of value, but there are serious problems associated to this concept. The main problem is that value is subjective and does not exist in the real world. Discussed in detail in [3, 12].

Yet, there are valuable aspects of the REA model. The approach in this paper is to represent REA in a better way.

2.2 DEMO Enterprise Ontology

DEMO is an engineering methodology to derive conceptual models of enterprises, based on an ontological theory, DEMO enterprise ontology (DEO) [1, 2]. DEO is comprised of four axioms and a theorem. DEMO is part of the emerging discipline of 'enterprise engineering' (EE) [2]. EE is founded on the same kind of theories as more mature engineering disciplines such as civil engineering, aviation and electronics. A claim for the quality of the applied methodology is guaranteed by the underlying theories, methodologies, formal methods [2, 5, 7] and a good body of empirical cases in many domains [5].

The DEMO methodology claims to provide models that meet the so-called C4-ness quality criteria [7]. Comprehensiveness refers to the condition that the model should encompass everything that is part of the ontology. This includes all concepts and relations of the ontology; nothing is missing.

Consistency refers to the absence of any anomalies of any kind. Conciseness refers to the requirement that anything that is not in the domain of the ontology should not be represented in any model. Coherence refers to the 'semantic meaningfulness of the symbols and their relations from every perspective'.

Specific results of C4-ness qualities are (i) that any enterprise that may exist in the real world, including virtual CC-CP enterprises, can be modeled correctly in one and only one way; and (ii) the DEMO model(s) for any such enterprise must provide concise and comprehensive factual knowledge about the operation of the enterprise. These two claimed results must be empirically tested for co-creation and co-production (Sect. 3). It is assumed, to be proven by validation and assessment, that expressing the REA ontology in DEMO may provide a DEMO model that is truthful and appropriate to represent the REA model and support accounting systems well.

3 The CC-CP Model

The purpose of the proposed CC-CP DEMO model [12, 13] is to be a generic specification of any financial or business interaction or transaction between our enterprise of interest and any external stakeholders such as customers, suppliers, personnel staff and taxation or other governmental institutions. In execution of that enterprise model factual knowledge must be provided for information systems. This model is claimed to capture any interactions between an enterprise and any stakeholder. It is a *generic pattern of interaction,* equivalent to the DEMO transaction.

3.1 Co-creation and Co-production Between an Enterprise and Its Stakeholders

Many highly specialized enterprises 'Contractors' do not have a well-defined portfolio of products with fixed prices but offer their capabilities to meet the specific requirements of their Principals. We define: co-creation captures the principal and the contractor(s) working together on the engineering of an acceptable artifact; co-production captures the shared production of the engineering artifact by both principal and contractor(s), including matching financial transactions.

In this paper, the original scope of co-creation and co-production has been extended to any stakeholder that interacts with our enterprise of interest; including customers, sub-contractors, suppliers, workers, tax offices etc.

It is assumed, but not proven and future research, that all – not only accounting systems – information systems that provide some appropriate perspective of the enterprise must be supported. The CC-CP fact model must specify all possible facts for any descriptive IS. To prove this completeness claim is future research.

3.2 Ontological Completeness Quality Criteria of the CC-CP Model

There are several mandatory quality and completeness quality criteria applicable. Missing criteria 1 or 2 for only one case renders the model worthless.

1. Completeness of the CC-CP model to capture any business interactions with any imaginable stakeholder in a truthful and appropriate way.
2. Completeness of the CC-CP model to capture any factual knowledge (Sect. 3.2) that maps to any REA concept for accounting systems.
3. Completeness of the CC-CP model to capture all factual knowledge that may be needed by any GAAP compliant accounting system. This is a wider quality criterion than requirement 2. This quality requirement is desirable but demands assessment of the common foundations of GAAP compliant accounting systems, which is future research.

3.3 The CC-CP Factual Information Support for Accounting Systems

The model must provide factual knowledge (informally "data about events") about the operation of the enterprise of interest for any imaginable information system – an Enterprise Information System or "EIS" - that provides some descriptive perspective of the operation of this enterprise. Key notion is that "facts", which are propositions about the world of phenomena, provide all required information for the descriptive accounting information systems.

The term factual knowledge refers to truthful propositions about phenomena in the world, In our case some phenomena in or about our enterprise of interest. In the FAR ontology [13] is specified that a fact is a proposition that may have a logic relation with other facts in a recursive way. A fact is a proposition that may have three values; true | false | undefined. While the meaning of the values true and false are clear, the value of "undefined" reflects the situation that for some unknown reason factual information is not available. In the FAR ontology there exist four kinds of facts:

1. Communicative facts; as defined by the DEMO transaction axiom.
2. Infologic and datalogic production facts. An example is the text of the contract of the CC-CP model. It is precisely the 'text only', without any actor commitments.
3. Facts about the world of phenomena not captured by the DEMO ontology, the kinds 1 and 2. Example: the exchange rate dollar – euro = 0.85. The value of this proposition can be true | false | undefined.
4. Any logic aggregated facts, or dependent facts, composed of logic relations (AND | OR | NOT relations) of other facts. Evaluation laws for the three-state logic.

The notion "full factual knowledge" is important. There are three completeness claims; (i) all interactions with any stakeholder with which there are transactions (financial or otherwise) must be captured well; (ii) for each interaction with a stakeholder all relevant facts must be provided; (iii) in addition there is the requirement that for each fact all relevant attributes of that fact must be provided.

Example: the fact represented by proposition "Person a is member of Club c" can be defined to be true if: The age of the proposed member is above 18 years of age; AND the membership admittance procedure has been approved; AND the membership fee has been paid. If these requirements have not all been met then the fact is not true. If any of the composing facts cannot be evaluated and returns the value undefined then the value of the fact becomes undefined. Relevant attributes of that fact may be the date when the membership became true, the duration of the membership. There are also relations to the person that is a member etc.

To summarize, the following propositions are formulated:

1. The CC-CP model captures any transaction based on business interactions betweenour enterprise of interest and any stakeholders, suppliers, subcontractors, staff etc.
2. Capturing of the business interaction between the enterprise of interest and stakeholders implies that CC-CP model provides a truthful and appropriate representation of all DEMO transactions.
3. For the enterprise of interest there exist a number of valuable descriptive perspectives – seen from the perspective of the stakeholders, shareholders, management etc. - of the operation of that enterprise. The relevant perspective of this paper is an accounting perspective provided by an EIS (enterprise information system), a REA compatible GAAP compliant accounting system.
4. There is a completeness claim for factual knowledge claims; all facts, and facts are complete with all attributes.

3.4   The CC-CP Fact Model

The proposed CC-CP Fact Model strictly follows the CC-CP Construction Model which is composed of six transactions, see Fig. 2. The presented Fact Model is described in three phases which correspond to the phases of the Construction Model and is illustrated in Fig. 3.

The co-creation phase includes T-1 and T-2 transactions. The object class CONTRACT which is the core concept in the whole CC-CP Fact Model is identified in this phase. The other object classes that are identified in this phase are PRODUCTION, PRICE, PRODUCTION-KIND, MONEY-KIND and the external object class ENTERPRISE. All mentioned object classes are primal classes, which means that they cannot be defined on the basis of other fact types. The lines between CONTRACT and ENTERPRISE labeled "principal of contract is enterprise" and "contractor of contract is enterprise" represent property types. Mandatory and uniqueness constraints indicate that a contract must have one enterprise as a principal and one different enterprise as a contractor.

The property type between the object classes CONTRACT and PRODUCTION indicates that each contract has only one production and each production has only one contract. The same holds for the property type between the object classes CONTRACT and PRICE.

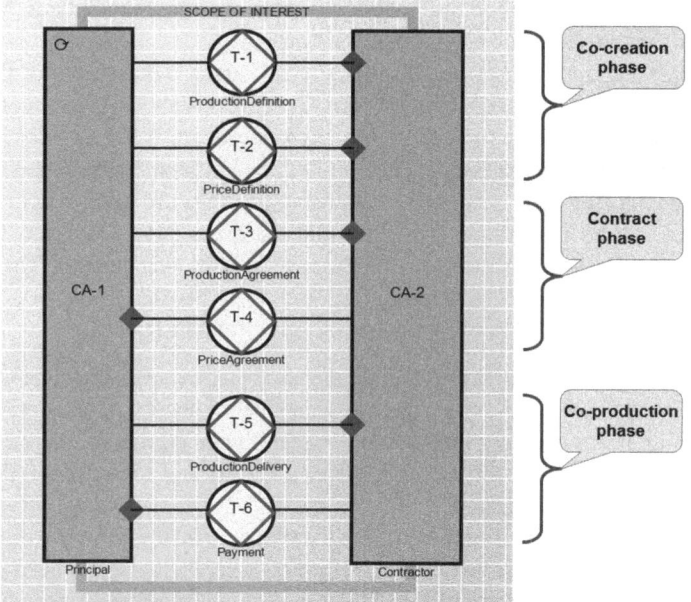

Fig. 2. The CC-CP Construction Model

The property type between the object classes PRODUCTION and PRODUCT-KIND expresses that one product can include more product-kinds which is in compliance with a purchase order containing more items. Each product-kind is further specified by value types which represent the volume (amount), the price per unit and the delivery day of the product-kind. The result kind "[production] was defined" is existentially independent unary fact kind which is the result of T-1 transaction.

The property type between the object classes PRICE and MONEY-KIND indicates that one price can have several money-kinds. Each money-kind is further specified by value types which represent the price of production and the day of payment. The result kind "[price] was defined" is existentially independent unary fact kind which is the result of T-2 transaction. From the implementation point of view it is supposed that T-1 transaction kind and T-2 transaction kind have each only one instance.

The contract phase includes T-3 and T-4 transactions. The result kind CONTRACT_SIGNED is another core concept and is a subclass of the object class CONTRACT. The figure illustrates that any contract can become a contract signed. This result kind becomes existent when T-3 is Promised and T-4 is Promised. From the above follows that in order to model a contract signing it is necessary to explicitly express two coordination facts and perform a logical aggregate over them. As the traditional DEMO methodology does not cope with this requirement, the FAR ontology was utilized to capture the above described task. From the implementation point of view it is supposed that T-3 transaction kind and T-4 transaction kind have each only one instance.

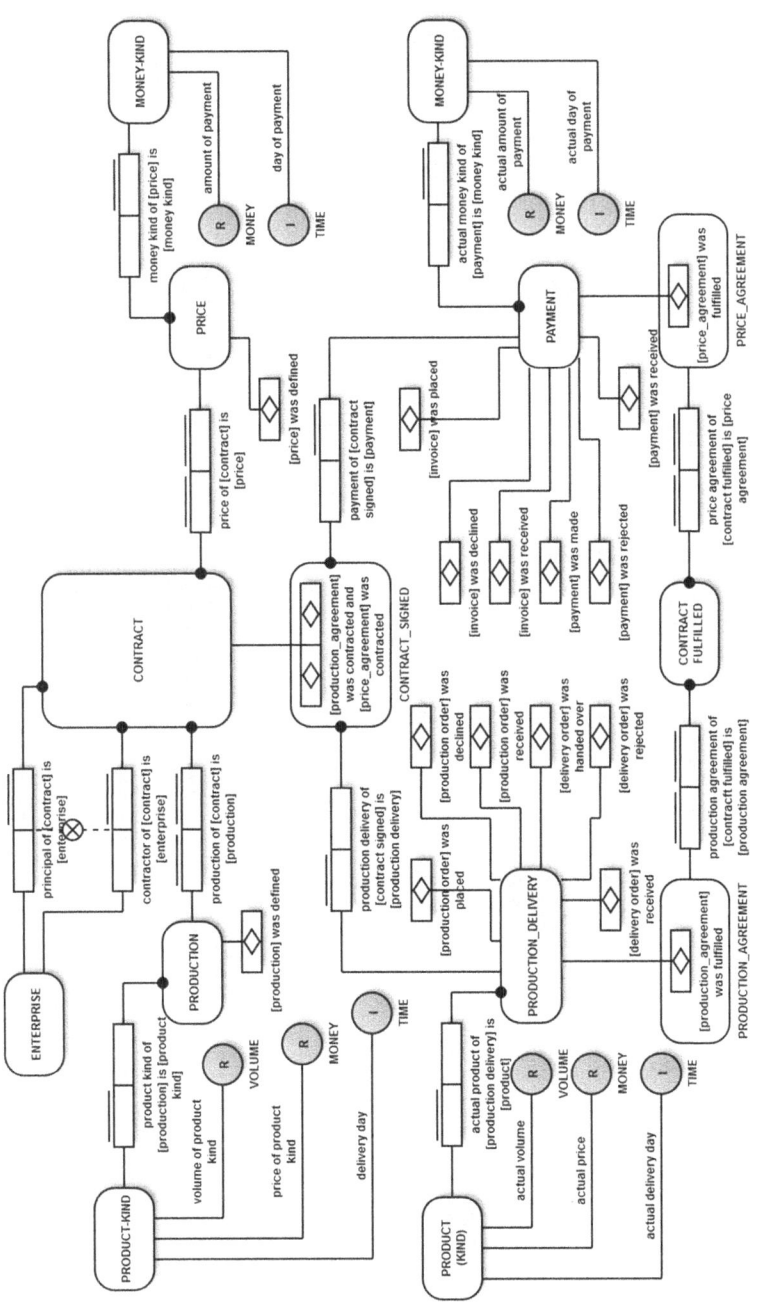

Fig. 3. The CC-CP Fact Model

The co-production phase is formed by T-5 and T-6 transactions and the execution and result phases of T-3 and T-4 transactions. The property type between the object classes CONTRACT_SIGNED and PRODUCTION_DELIVERY indicates that one contract_signed can have more production deliveries which is in compliance with the modeling reality. From the implementation point of view it is supposed that T-5 transaction can have one or more instances.

The property type between the object classes PRODUCTION_DELIVERY and PRODUCT (KIND) expresses that one production_delivery can have more products. Each product is further specified by value types which represent the actual volume, the actual price per unit and the actual delivery day. A product, as such, can be identifiable or quantifiable or both. In case a product is identifiable, it can have a serial number and the notion of product can be used. If the product is only quantifiable the notion of product kind is used. The result kind "[production] delivery" is existentially independent unary fact kind, which is the result of T-5 transaction.

The property type between the object classes PAYMENT and MONEY-KIND indicates that one payment can represents more money-kind which is in compliance with a payment order containing more money kinds. Each MONEY-KIND is further specified by value types which represent the actual price of production and the actual day of payment. The result kind "[payment] was made" is existentially independent unary fact which is the result of T-6 transaction.

The result kind PRODUCTION_AGREEMENT is a subclass of the object class PRODUCTION_DELIVERY. It means that the object class PRODUCTION DELIVERY can become the result kind PRODUCTION_AGREEMENT when the fact "production_agreement was fulfilled" becomes existent. The result kind PRICE_AGREEMENT is a subclass of the object class PAYMENT. It means that the object class PAYMENT can become the result kind PRICE_AGREEMENT when the fact "price_agreement was fulfilled" comes into existence.

The object class CONTRACT_FULFILLED becomes existent when the result types PRODUCTION_AGREEMENT and PRICE_AGREEMENT come into existence. The meaning of this object class is that obligations concerning production_agreement and price_agreement as declared in the object class CONTRACT_SIGNED were fulfilled and the contract is completed. The object class CONTRACT_FULFILLED represents the duality relationship in REA.

3.5 Conceptual Mapping of the CC-CP FACT Model to REA Model Concepts

Despite the fact that the conceptual mapping is rather simple and needs further rigorous elaboration, it captures the core issue. The DEMO Bank Contents Table, which is shown in Table 1, contains object classes, fact types, and transaction banks, in which their instances are contained.

Table 1. The Bank Contents Table of the CC-CP Model

Bank	Independent/Dependent fact
T1	CONTRACT
	the principal **of** Contract
	the contractor **of** Contract
	ENTERPRISE
	the production **of** Contract
	PRODUCTION
	the product-kind of Production
	PRODUCT_KIND
	the volume **of** Product-Kind
	the price **of** Product-Kind
	the delivery day **of** Product-Kind
	the production **of** Contract **is** defined P1
T2	PRICE
	the price **of** Contract
	MONEY_KIND
	the amount **of** payment **of** Money-Kind
	the day **of** payment **of** Money-Kind
	the price **of** Contract **is** defined P2
T3	**the** production_agreement **is** fulfilled P3
T4	**the** price_agreement **is** fulfilled P4
T3, T4	**the** production_agreement **is** promised **and** **the** price_agreement **is** promised
	the production requisition
	CONTRACT_FULFILLED
	the production_agreement **is** fulfilled **and** **the** price_agreement **is** fulfilled P3 and P4
T5	PRODUCTION_DELIVERY
	the production delivery **of** Contract_Signed
	the product **of** Production_Delivery
	PRODUCT
	the actual volume **of** Product
	the actual price **of** Product (price per unit)
	the actual delivery day **of** Product
	the production order placed (sent) T5.rq
	the production order declined T5.dc
	the production order received T5.pm
	the delivery order handed over T5.st
	the delivery order receipt T5.ac (P5)
	the delivery order rejected T5.rj
T6	PAYMENT
	the payment **of** Contract_Signed
	the money-kind **of** Payment
	MONEY_KIND
	the actual amount **of** payment **of** Money-Kind
	the actual day **of** payment **of** Money-Kind
	the invoice placed (sent) T6.rq
	the invoice declined T6.dc
	the invoice received T6.pm
	the payment made (sent) T6.st
	the payment receipt T6.ac (P6)
	the payment rejected – dispute T6.rj

The coordination fact "the delivery order receipt" means that the production was not only delivered but was also accepted by the principal. At this time, the corresponding production fact comes into existence. The same holds for the coordination fact "the payment receipt" which means that the payment was not only sent by the principal but was also accepted by the contractor. At the same time the corresponding production fact becomes existent.

The conceptual mapping deals with the DEMO CC-CP model fact kinds and their mapping to REA concepts and relationships as follows. The production fact "the production of Contract is defined" which becomes existent as a result of T-1 transaction contains all dependent facts (property types, attributes types) that are needed for one kind (decrement/increment) of an REA commitment. "The price of Contract is defined" is the next production fact which comes into existence as a result of T-2 transaction. The T-2 transaction instance contains all dependent facts (property types and attribute types) that are needed for one kind of an REA commitment. The aggregate coordination fact "the production of Contract is promised and the price of Contract is promised" is mapped into the reciprocity relationship that relates a different kinds of commitments to each other. The commitments are related to the corresponding resource types and economic agent types. "The production requisition" is a dependent fact type, which is mapped into the reservation relationship in the REA model.

The number of instances of the T-5 and T-6 transaction types corresponds to a number of production deliveries and a number of installments, respectively. The T-5 transaction instance captures one production delivery, which is in compliance with reality. The independent production fact "the delivery order receipt" is accompanied by the dependent facts of the property types and attributes types. From the accounting perspective the most important are explicitly expressed coordination facts that capture the necessary inventory system events. The CC-CP model is able to register all these events.

The T-6 transaction instance captures one payment (installment) in compliance with reality. The independent production fact "the payment receipt" is accompanied by dependent facts of the property types and attributes types. From the accounting perspective, explicitly expressed coordination facts that capture the necessary accounting system events are the most important. These events are: *sending an invoice, receiving an invoice, making a payment*. The T-5 and T-6 transaction instances can provide coordination facts of decline and reject. Their practical meaning is as follows. "The production order was declined" or "the delivery order was rejected" in case of the T-5 transaction instance, and "the invoice was declined" or "the payment was rejected" in case of the T-6 transaction instance.

To summarize the following results are found. From the above simple analysis follows that the CC-CP model provides all the facts that are necessary for the REA exchange model. The decline and reject coordination facts have no equivalent in the REA model but have equivalents in reality. In addition, the CC-CP model captures more precisely and truthfully the facts that pertain to the signing of a contract and the facts that concern the fulfilling of a contract.

Based on these simple and not rigorous assessments it is claimed that the DEMO CC-CP model fully captures the facts needed by the REA model.

4 Benefits of the REA Model Represented by the DEMO CC-CP Model

The following benefits are provided by the CC-CP model and DEMO [1]:

1. The CC-CP model is extensible without loss of its capabilities. Supporting transactions can be added to provide more control of the enterprise operation. Example: An employee is permitted to send a quotation for an order to a customer, a legally binding commitment, but must have first an approval from a colleague. This is an imposed business rule that must be enforced. To model this correctly, a transaction must be created between the employee and the colleague. The production fact of that transaction is an approval or a rejection. A business rules inhibits the c-act to send the quotation until that permission Pfact becomes true – approved. If the Pfact is rejected it will be never possible to send the quotation. DEMO enables precise definition and execution of these kinds of rules [13].
2. The provision of all historic events, all documents, all commitments, with time/date stamps, with guaranteed completeness in case of a dispute. This is also a complete litigation case file. By applying the blockchain technology, the case file becomes absolute trustworthy, it will be impossible to modify it.
3. The model can be extended or refined for any imaginable specific business situation and adding defined business rules [13]. Including partially accepted deliveries, return deliveries, not accepted payments, transaction roll-back etc. These claims are promising but unproven benefits and can be considered more as a topic of future research.
4. The model must be free from anomalies such as deadlocks. While in the real world it is possible to devise business rules may create anomalies such as a deadlock, a deadlock condition can be modeled also. Though undesirable, it must be possible to implement some system with a deadlock. Model simulation and validation identifies and mitigates anomalies such as deadlock and other anomalies.

5 REA Model-Driven GAAP Compliant Systems and Theories

New ontological theories promise a model-driven approach for the development of GAAP compliant accounting systems. The development of some GAAP compliant accounting system is then simplified to devising a conceptual model, expressed in a GAAP language, typically done by accounting experts. This model with matching software engine constitutes directly the GAAP compliant accounting system, which eliminates programming to a large degree (future research).

The theoretical foundations of the proposed approach are briefly described:

Guizzardi [4] proposed the foundations of ontological theories and a framework. This framework captures (i) the phenomena of a specific domain in the real world; (ii) the corresponding conceptualizations and (iii) an ontological modeling language. Any proposition expressed in that ontological modeling language specifies some phenomena that (may) exist in that domain in the real world.

Dietz J.L.G. [1] provided the DEO, DEMO Enterprise Ontology, a domain ontology that captures any enterprise that operates in the real world. The DEMO methodology provides conceptual models, formal representations of enterprises. Dietz J.L.G. provided also the Generic Systems Development Methodology [GSDP].

Van Kervel [6] extended the Guizzardi framework for static ontologies also for dynamic ontologies and for a model executing software engine. This is based on the GSDP methodology and results in the Generic Systems Development Process for Model-Driven Engineering [GSDP-MDE] of (software) systems. This approach has been proven; the DEMO engine has been built in this way [7].

The benefits are that in this way the development of a GAAP compliant accounting system demands much less resources; "only" a conceptual model is needed (best case). Also the validation that the accounting system is GAAP compliant is much easier. In case the GAAP rules change, the model can be changed very quickly. The automatic integration of different GAAP compliant systems to one coherent representation is another promise.

6   Conclusion and Results

It has been shown that he CC-CP Fact Model contains all required facts, with proper fact mapping for REA accounting systems, plus transactional behavior such as reject delivery, decline order, reject payment etc. The complete and correct factual mapping shows that the CC-CP model is appropriate to serve REA accounting systems.

However, much future research is needed to validate our generally careful claims: (i) more rigorous assessment of conceptual alignment REA - DEMO concepts; (ii) more empirical appropriateness case studies to support the claim that the CC-CP model captures any enterprise - enterprise co-creation and co-construction operation; (iii) in this perspective, many implementation-specific extensions of the CC-CP model; (iv) progress in the application of the GSDP-MDE approach and in conceptual modeling; the fact that one application - the DEMO engine - works well does not guarantee its generic applicability; (v) Notably conceptual modeling of GAAP compliant systems is a new domain.

Acknowledgements. The paper was supported by the grant provided by Ministry of Education, Youth and Sports Czech Republic, reference no. SGS09/PRF/2017.

References

1. Dietz, J.L.G.: Enterprise Ontology: Theory and Methodology. Springer, Heidelberg (2006)
2. Dietz, J.L.G., Hoogervorst, J.A.P.: The discipline of enterprise engineering. Int. J. Organisational Des. Eng. **3**(1), 86–114 (2013)
3. Dietz, J., Aveiro, D., Pombinho, J., Hoogervorst, J.: An ontology for the τ-theory of enterprise engineering. Frontiers Artif. Intell. Appl. **267**, 386–395 (2014)
4. Guizzardi, G.: Ontological foundation for structural conceptual models. Ph.D. theses, University of Twente (2005)
5. van Kervel, S.J.H., Dietz, J.L.G., Hintzen, J., van Meeuwen, T., Zijlstra, B.: Enterprise ontology driven software engineering. In: Proceedings of International Conference on Software Paradigm Trends (2012)

6. van Kervel, S.J.H.: Ontology driven enterprise information systems engineering: Ph.D. thesis, University of Technology Delft (2012)
7. Dudok, E., Guerreiro, S., Babkin, E., Pergl, R., Kervel, S.J.H.: Enterprise operational analysis using DEMO and the enterprise operating system. In: Aveiro, D., Pergl, R., Valenta, M. (eds.) EEWC 2015. LNBIP, vol. 211, pp. 3–18. Springer, Cham (2015). doi:10.1007/978-3-319-19297-0_1
8. Dunn, C.L., Cherrington, O.J., Hollander, A.S.: Enterprise Information Systems: A Pattern Based Approach. McGraw-Hill/Irwin, New York (2004)
9. Hruby, P.: Model-Driven Design Using Business Patterns. Springer, Heidelberg (2006)
10. McCarthy, W.E.: The REA accounting model: a generalized framework for accounting systems in a shared data environment. Account. Rev. **57**, 554–578 (1982)
11. Hunka, F., Zacek, J.: Detailed analysis of REA ontology. In: Aveiro, D., Tribolet, J., Gouveia, D. (eds.) EEWC 2014. LNBIP, vol. 174, pp. 61–75. Springer, Cham (2014). doi: 10.1007/978-3-319-06505-2_5
12. Hunka, F., Kervel, S.J.H., Matula, J.: Towards co-creation and co-production in production chains modeled in DEMO with REA support. In: Aveiro, D., Pergl, R., Gouveia, D. (eds.) EEWC 2016. LNBIP, vol. 252, pp. 54–68. Springer, Cham (2016). doi:10.1007/978-3-319-39567-8_4
13. Skotnica, M., van Kervel, S.J.H., Pergl, R.: Ontological foundation for the software executable DEMO action and fact models. In: Aveiro, D., Pergl, R., Gouveia, D. (eds.) EEWC 2016, LNBIP, vol. 252, pp. 151–165. Springer, Heidelberg (2016)
14. Nuffel, D., Mulder, H., Kervel, S.: Enhancing the formal foundations of BPMN by enterprise ontology. In: Albani, A., Barjis, J., Dietz, Jan L.G. (eds.) CIAO!/EOMAS -2009. LNBIP, vol. 34, pp. 115–129. Springer, Heidelberg (2009). doi:10.1007/978-3-642-01915-9_9

SysPRE - Systematized Process for Requirements Engineering

Ana Neto[1(✉)], Duarte Pinto[1], and David Aveiro[1,2]

[1] Faculty of Exact Sciences and Engineering, University of Madeira,
Caminho da Penteada, 9020-105 Funchal, Portugal
ana.b.neto@gmail.com, duarte.pinto.oelabuma@gmail.com
[2] Madeira Interactive Technologies Institute, Caminho da Penteada, 9020-105 Funchal, Portugal
daveiro@uma.pt

Abstract. The domain of Knowledge Discovery (KD) and Data Mining (DM) is of growing importance in a time where more and more data is produced and knowledge is one of the most precious assets.

Having explored both the existing underlying theory, the results of the ongoing research in academia and the industry practices in the domain of KD and DM, it was found that this is a domain that still lacks some systematization.

It was also noticed that this systematization exists to a greater degree in the Software Engineering and Requirements Engineering domains, probably due to being more mature areas.

In this paper we propose SysPRE - Systematized Process for Requirements Engineering in KD projects to systematize the requirements engineering process for these projects so that the participation of enterprise stakeholders in the requirements engineering for KD projects can increase.

Keywords: Knowledge discovery · Data mining · Requirements engineering · DEMO

1 Introduction

Software development has been around for several decades now and discussion on its failures and successes has been strong.

It all started with the Standish Group's Chaos Report of 1994 [1] that stated that projects that did not meet customer satisfaction and/or went over time or budget in a significant way corresponded to 53%. It was a bit shocking to see a figure that amounted for over half of the projects and a discussion about a software crisis was started.

This report, however, was since then criticized for lack of peer review, for not having a complete description of the study design or of the project selecting criteria, and for defining successful and failed projects in a way that may bias the study [2, 3, 5].

Over 20 years later, the debate is still on, but there seems to be an agreement on the failure rate of software development projects having dropped [5–7]. Although the values do not coincide, they show a decrease tendency that may be significant if you take into account that projects are increasingly complex.

© Springer International Publishing AG 2017
D. Aveiro et al. (Eds.): EEWC 2017, LNBIP 284, pp. 166–180, 2017.
DOI: 10.1007/978-3-319-57955-9_13

One of the areas of software development that has helped this increased success in software projects is Requirements Engineering (RE), following previous research such as [8, 9]. Furthermore, according to [6], one of the three main reasons for the positive development is that the communication of requirements has much improved. [10] makes an even stronger statement that "Meets user requirements" is the most important success criteria for both users (96%) and project managers (81%).

Knowledge discovery and data mining are much more recent areas than software development and less mature fields. For instance, if considering process model development to be a sign of maturity, it can be seen that the first process model for this area dates back to 1996 [11], while in software development, the well-known Waterfall model goes back to 1970 [12].

Nonetheless, it is indisputable that knowledge discovery and data mining are of growing importance in a time where more and more data is produced.

Data production numbers are, in fact, staggering, for example, 144.000 h of video are uploaded to YouTube per day [13], 182.900.000.000 emails are sent per day [14] and 1.000.000.000 pieces of content are shared on Facebook per day [15].

This results in massive amounts of data. Facebook has one of the largest data warehouses in the world, storing more than 300 petabytes [16].

With such a large production of data and in a time when knowledge is one of the most precious assets, it is no wonder that knowledge discovery and data mining are of increasing importance.

The road for knowledge discovery and data mining projects is to increase systematization, as the area becomes more main stream.

This seems important because the trends in this area, currently, are to have larger projects (with larger amounts of data involved) and, at the same time, to have the people involved in those same projects with lower technical skills and very little time to experiment with different approaches [17].

Within the knowledge discovery projects, the area of requirements engineering is the one that can reap more benefits thanks to a higher level of systematization.

Firstly, because requirements engineering is particularly neglected in this type of projects. Some authors even argue that this type of projects should be based on the available data and not on stakeholders' requirements [18].

Secondly, because, being a less mature field, less systematization efforts have been made so far and when they occur, the participation of enterprise stakeholders will be improved and facilitated and the area will follow software engineering in general, that has improved in terms of customer satisfaction and time and budget compliance.

For these reasons, the research question was "How can systematization be brought into Knowledge Discovery projects, in general, and into their Requirement Engineering phase, in particular, aiming at improvements in their success rate?"

The research started by analysing the Knowledge Discovery process through a systematic review of the state-of-the-art in academia and industry regarding knowledge discovery and data mining process models. To conclude this review a comparing of the main process models found was made.

Then the Requirements engineering area was analysed in a similar way followed byocusing on requirements engineering for KD projects. It was found that requirements engineering for KD is different. That is why it is claimed here that a requirements engineering for KD process model is needed and SysPRE, a Systematized Process for Requirements Engineering designed specifically for KD projects is proposed.

SysPRE, began from an initial textual description which was then formally specified as a DEMO ontology [45]. This formal specification was instantiated in two case studies so that trivial and non-trivial errors could be identified and the necessary adjustments made.

SysPRE synthesises the knowledge obtained for the state-of-the-art reviews in a way that can be helpful for enterprises and other organizations with KD projects both for novice and expert users, with the hope of bringing improvements to the success rate of such projects.

2   Knowledge Discovery Process and Demo Specification

In this section the Knowledge Discovery Process (KDP) will be described as seen after analysing the existing process models listed in Sects. 2.1 and 2.2 with special detail with what regards Requirements Engineering within the KDP.

This specifically considers business KDPs, but this description would also be accurate for other types of organizations, namely governmental or non-profit.

2.1   Knowledge Discovery

The need for a process model stems from the fact that data mining is non-trivial. In 2006, Bernstein et al. referred that "there are many possible choices for each stage, and only some combinations are valid. Because of the large space and nontrivial interactions, both novices and data mining specialists need assistance" [19].

Still the need for a process model goes back to 1989, when it was first discussed during the IJCAI workshop on Knowledge Discovery in Databases (KDD) [20]. This was the original workshop that started the series of KDD workshops that, from 1995 onwards, grew into KDD conferences. Still, only in 1996 the first model was formally proposed.

This original KDD model consisted in nine steps: learning the application domain, understanding the domain and any relevant prior knowledge but also identifying the goal of the process; creating a target dataset; data cleaning and pre-processing; data reduction and projection; function of data mining selection (e.g., summarization, clustering); data mining algorithm(s) selection and specification of relevant parameters; data mining, which means the actual search for patterns; interpretation of the results; using discovered knowledge, which could be done in many ways, such as incorporating the knowledge into another system or simply generating a report of the findings.

From this model other models derived such as Ganesh et al. [21] and Adriaans and Zantinge [22] in 1996, Brachman and Anand [23] in 1997, Berry and Linoff [24], Cabena et al. [25], Knowledge Discovery Life Cycle (KDLC) model by Lee and Kerschberg [26] 1998 or Buchner et al. [27] in 1999.

The most widely used in the industry however was CRISP-DM [46]. Created in 1997 by a group of organizations involved in data mining (NCR, SPSS, Daimler-Chrysler and

OHRA). The first version was published in August 2000 [28]. Between 2006 and 2008 there were efforts to launch a second version of CRISP-DM, which was referred to as CRISP-DM 2.0, but no result was ever published.

The CRISP-DM model life cycle consists of six iterative steps: business understanding; data understanding; data preparation; modelling; evaluation; deploying.

To CRISP-DM many variations were proposed over the years, such as Rapid Collaborative Data Mining System (RAMSYS) model [31] in 2001, Data Mining for Industrial Engineering (DMIE) by Solarte [32] in 2002, Data Mining and Knowledge Discovery (DMKD) model by Cios and Kurgan [33] in 2005, Ontology Driven Knowledge Discovery (ODKD) by Gottgtroy [34] in 2007, Knowledge and Discovery and Communication Framework (KDCF) by Rennolls and AL-Shawabkeh [35] and ASD-DM by Alnoukari et al. [36] in 2008 or IKDDM by Osei-Bryson [37] in 2012.

Other models include Catalyst methodology in 2003 [30]. This methodology has two parts: business modelling and data mining. For each part, a detailed step-by-step methodology is suggested. Originally it was proposed both in printed form and online, and both formats followed a hyperlink structure.

Considering both parts of the methodology as a whole, we can say that it has six steps: business modelling; data preparation; tool selection; mining; refining; deploying.

What makes this methodology interesting is the level of detail that is includes in each step. It is very focused on what needs to be done and how it can be done. This is organized in what the author calls "boxes". There are four types of "boxes": Action Boxes, Discovery Boxes, Technique Boxes, and Example Boxes.

And finally SEMMA was created to be used is a specific application, SAS Enterprise Miner [29].

The acronym SEMMA stands for sample, explore, modify, model, assess, which are basically the five iterative steps proposed: sample, which consists of extracting sample data (optional step); explore, which means the exploring the data or the sample data in order to be able to simplify the model; modify, which can include any cleaning, preprocessing, reductions or projections deemed necessary; model, which is the actual search for patterns; assess, which is the evaluation and interpretation of the results.

SEMMA however is tied to the SAS Enterprise Miner tool and therefore overlooks any steps that are not related to the tool, namely any business understanding tasks.

2.2   Requirements Engineering

The IEEE Standard Glossary of Software Engineering Technology [38] defines a software requirement as:

1. A condition or capability needed by a user to solve a problem or achieve an objective.
2. A condition or capability that must be met or possessed by a system or system component to satisfy a contract, standard, specification, or other formally imposed document.
3. A documented representation of a condition or capability as in 1 or 2.

In short, a software requirement is something that we expect the software to meet.

In the studied methods there was a special focus on six, Waterfall by Winston Royce [12] in 1970, Spiral by Barry Boehm [39] in 1986, Rapid Application Development (RAD) by the New York Telephone Company in mid-1970s, becoming notorious in the early 90's by James Martin and his approach [40], Rational Unified Process (RUP) by the Rational Software Division of IBM [41], Agile proposed in 2001 in the Agile Manifesto [42] and Goal-Oriented Requirements Engineering (GORE).

2.3 PIF and CAP Analysis

To the KDP a Performa-Informa-Forma (PIF) analysis and a Coordination-Actors-Production (CAP) analysis were made with the goal to gain insight to what concepts and activities are important in the KD process. Namely, in terms of activities, the Performa items are the truly relevant ones and will later be the transactions of the DEMO specification of SysPRE.

Most of the Performa-Informa-Forma is being omitted remaining only the Performa items in italic. The Coordination-Actors-Production analysis was done simultaneously by enclosing a piece of text indicating an actor role between the brackets "[" and "]". Transaction's id (for instance T01) are also marked next to Performa items.

The knowledge discovery process begins {T01} when the [business analyst] *realizes that there is a business problem or opportunity* {T02} in which Knowledge Discovery and Data Mining might be helpful. More commonly, the [business analyst] starts with a question and needs certain information relevant to the decision he must make.

He or she starts by trying to *learn* {T03} as much as possible about the business and the application domain. He will *identify* the [stakeholders] {T04}. He will try to *understand what issues are important* for the [stakeholders] {T05}. The five core issues are [30]: product (goods or services, tangible or intangible); place; price; time; quantity.

The [business analyst] will *classify the knowledge discovery process* as {T06}:

- Demand driven - process is aimed to fulfil the information requirements of the users
- Data driven - process is aimed to discover the best use to the specific existing data
- Exploratory - process is designed to find how KD and DM in general can offer value within that specific business

He will try to discover any relevant prior knowledge, namely the currently existing solutions for the problem, and *identify the goal* for the project {T05}.

If it is an exploratory process, the [business analyst] will *identify several possible goals* {T05} and *review his stakeholders' identification* {T04} for each one (including the core issues that each one might be concerned with {T05}).

Since starting the project might have costs, the [business analyst] might have to ask for *approval* {T14} for the data mining project to the [business manager]. The [business manager] might *ask* {T13} the [project manager] for a cost and resources estimation so that he can *decide on the approval* {T14}. The [project manager] *will create the cost, time and resources estimates or a project plan* {T13}, if necessary. The [project manager] will hand these to the [business manager]. The [business manager] will *decide to go ahead or not* {T14}, that is, he will decide on the feasibility of the KD project. If the decision is to go ahead, the [project manager] might have to get the resources (human or otherwise) that are necessary and that were not available in the beginning.

If it is a demand driven project, the [business analyst] will then begin *eliciting specific requirements* {T07}. If it is a data driven project, the [business analyst] will then proceed by asking the [data analyst] to perform the data analysis. A hybrid approach is also possible, in which both will happen in parallel. For the *requirements elicitation* {T07}, the [business analyst] will *choose the elicitation techniques* {T08}, which might be one or more. He will execute them and document the resulting requirements from each technique at what is judged to be an appropriate level of detail. These requirements will be mostly information demand requirements, that is, requirements that describe why and how the [stakeholders] need specific information. The [business analyst] will also *elicit non-functional requirements* {T07}, and for that he will be particularly concerned with the delivery mechanism (how will the results be physically made available to the [end user]? What tools will the [user] employ to view it?), the format (will the [user] view the results in reports, dashboards, or other formats?) and the degree of interaction needed (to what extent must the [user] be able to manipulate the results following delivery?).

A detailed analysis of the requirements will be done by the [business analyst]. The [business analyst] and the multiple [stakeholders] will *negotiate* to:

- *Decide which requirements are accepted* {T09} (which, in fact, is the same as deciding the system boundaries or scope)
- *Do a triage and prioritization of the requirements* {T10}
- *Assess requirements risks* {T11}

The [business analyst] will *validate* {T12}, that is, check for completeness and for consistency the resulting requirements.

The *triage and prioritization* {T10} should be done after the *validation* {T12}, as the validation {T12} process might result in adding, changing or removing some requirements.

The [business analyst] will also need data, so he will ask the [data analyst]. Again, note that in a demand driven project this request will normally happen after the requirement elicitation {T07}, but in a supply driven project the data gathering that we will describe next will happen before the *requirement elicitation* {T07}. The [data analyst] will *look for the raw data* {T15} to use for the project. The data might come from databases, internal or external, or from other sources. It might also need still to be collected for this specific purpose. The [data analyst] will need to *select the data* {T16} and *decide if and when the data might need to be combined* {T17}. If the [data analyst] considers the data to be too large for an initial analysis, he might *consider using a sample* {T17} of the data.

The [data analyst] will also try to understand the data. To begin with, if the data was already available at the beginning of the project, the [data analyst] should find the business motivation to collect and store the data in the first place, as it might provide some insights. From the data understanding he might *suggest a possible hypotheses or objective* {T18} to the [business analyst]. He might also *identify constrains* {T19} that arise from the data, so he will inform the [business analyst] of the detected constrains.

Since the raw data might be incomplete, noisy or inconsistent, the [data engineer] will perform *data cleaning, pre-processing and transformation* {T17}. This might include filling missing values, normalization, discretization, reduction, projection or other techniques. The data cleaning, pre-processing and transformation is guided by the data itself and also by what data mining techniques are going to be used on the data. The [data miner] *selects the tool* {T20} to be used (for the same project, more than one tool might be used). For selecting the tool he will start by *identifying possible tools* {T28} and *decide on how he will compare them* {T20}, specifying the evaluation criteria that are important and how the evaluation will be performed (for instance, he might decide to run a specific algorithm using all the tools and a sample of the data). He will then proceed with the evaluation and *choose the tool* {T20} (or tools). The [data miner] also *selects the data mining technique* {T21} (e.g., summarization, classification, regression, clustering) and the *specific algorithms* {T22}. For the same project, more than one tool might be used, as well as more than one data mining technique and one algorithm.

Some authors believe the choice of data mining technique can be simplified *to four decisions* {T21}.

The [data miner] will entail the prepared data to the tool and be responsible for the *generation of the model* {T24}. This means he will have, for instance, to *decide on the appropriate parameters* {T23}.

After the actual data mining has occurred and the KD results are available, both the [domain expert] and the [strategic manager] will *analyse the results* {T25}.

The [domain expert] *analyses* {T25} the data mining result, in the sense that he *evaluates how the results fit his domain knowledge* {T25}, possibly resulting in the need for refining what was done previously through:

- *Creating new questions or hypothesis* {T18} for the [business analyst]
- *Pointing the need for new or more data* {T15} for the [data analyst]
- *Indicating the need to use a different function* {T21} *or algorithm* {T22} *or simply to adjust parameters* {T23} to the [data miner]

The [strategic manager] *interprets and evaluates* {T25} the data mining result, in the sense that he *evaluates how these results are relevant to or have an impact* {T25} on the current or future business situation.

The [knowledge engineer] will use the analysis results from the [domain expert] and the [strategic manager] and make sure the discovered knowledge is used. He will *specify* {T26} how the knowledge discovery result should be deployed, for instance he can decide that an annual report should be produced for the senior management. The knowledge discovery result will then be deployed to the [end users] as planned.

2.4 Transaction Result Table

From the Performa-Informa-Forma analysis and Coordination-Actors-Production analysis the Transaction Result Table (TRT) was the following.

This table shows the transactions (that correspond to the main tasks of the process) and the result types corresponding to each transaction. In the result types, we can see (between square brackets) the main concept that is being created or whose state is being changed.

The last transactions (T28 to T31) refer to the specification of an elicitation technique for requirements or, regarding the data mining stage, the specification of a tool, data

Table 1. Transaction Result Table

Transaction		Result type	
Id	Name	Id	Description
T01	Knowledge Discovery	R01	[knowledge discovery process] was realized
T02	Problem/Opportunity identification	R02	[problem/opportunity] was identified
T03	Problem/Opportunity analysis	R03	[problem/opportunity] was analysed
T04	Stakeholder identification	R04	[stakeholder] was identified
T05	Goal/core issue identification	R05	[goal/core issue] was identified
T06	Process classification	R06	[knowledge discovery process] was classified
T07	Requirement elicitation	R07	[requirement] was elicited
T08	Choice of elicitation technique	R08	[elicitation technique] was chosen
T09	Decision of scope	R09	decision on whether the [requirement] is in scope was made
T10	Requirement prioritization	R10	priority of [requirement] was defined
T11	Assessment of requirement risks	R11	risks of [requirement] were assessed
T12	Requirement validation	R12	[requirement] was validated
T13	Cost and resources estimation	R13	[cost and resources] were estimated
T14	Go-no-go Decision	R14	go-no-go decision of [knowledge discovery process] was made
T15	Data source identification	R15	[data source] was identified
T16	Data selection	R16	[data] was selected
T17	Data preparation	R17	[data] was prepared
T18	Hypothesis creation	R18	[hypothesis] was created
T19	Data constrain identification	R19	[data constrain] was identified
T20	Choice of tool	R20	tool was chosen for [result]
T21	Choice of data mining technique	R21	data mining technique was chosen for [result]
T22	Choice of algorithm	R22	algorithm was chosen for [result]
T23	Choice of data mining parameter	R23	data mining parameter was chosen for [result]
T24	Result obtention	R24	[result] was obtained
T25	Result analysis	R25	[result] was analysed
T26	Deployment specification	R26	[deployment] was specified
T27	KD area artefact management	R27	[KD area artefact] was managed
T28	Elicitation technique specification	R28	[elicitation technique] was specified
T29	Tool specification	R29	[tool] was specified
T30	Data mining technique specification	R30	[data mining technique] was specified
T31	Algorithm specification	R31	[algorithm] was specified
T32	Data mining parameter specification	R32	[data mining parameter] was specified

mining technique, algorithm or data mining parameter that was previously unknown to the system. This is necessary as the knowledge discovery and data mining area is very dynamic and it is very likely that new tools, data mining techniques, algorithms or data mining parameters need to be considered.

T27 is the transaction that manages all this. The elicitation techniques, tools, data mining techniques, algorithms and data mining parameters are referred to as artefacts in the context of T27 (KD area artefact management) (Table 1).

2.5   Object Fact Diagram

Due to space constrains the DEMO's Actor Transaction Diagram and the Process step diagram are omitted in this paper.

We then specified the DEMO's Object Fact Diagram (OFD).

In this diagram it can be seen the classes that correspond to the main concepts identified in the DEMO transactions of the Transaction Result Table, as well as other related classes, the fact types that are associated with each class and the cardinalities and dependence laws.

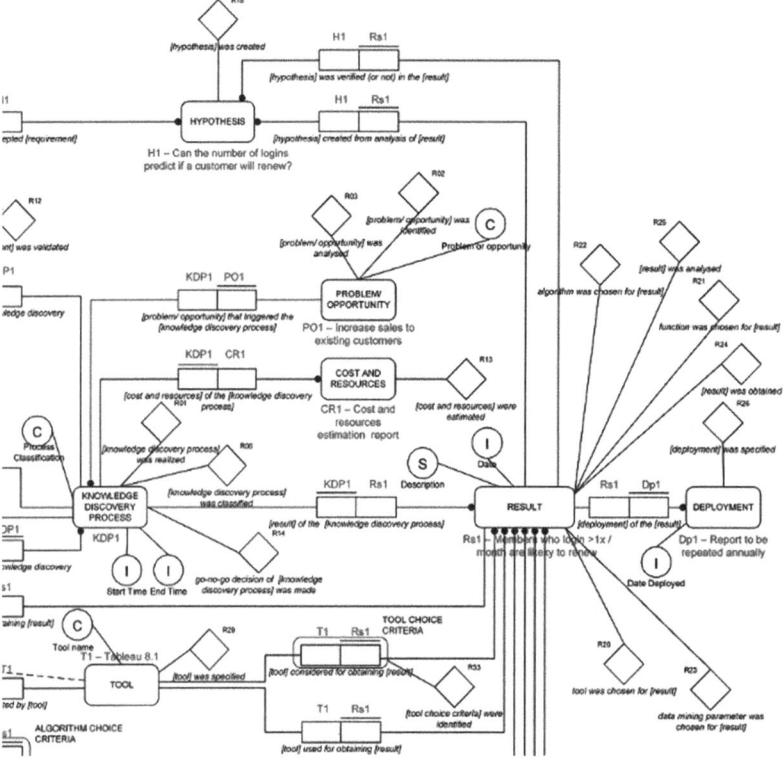

Fig. 1. Object Fact Diagram (Part 1)

In the image marked in red are comments of an instantiation of each class derived from a concrete case of a real organization, so that the interpretation of the diagram is easier (Fig. 1).

The main class of this OFD is the KNOWLEDGE DISCOVERY PROCESS (KDP), related to the main transaction T01. Each instance of this class will specify a particular KDP. Most of the classes that follow (in all caps text) are self-explanatory, so will presented as the example is described.

Instances of the class PROBLEM/OPPORTUNITY specify a problem or an opportunity that triggered the KDP. Let's say that a company wants to increase its sales to existing customers. The company we are considering, sells memberships, so basically they'll want to increase the percentage of customers that renew their memberships. This is the problem/opportunity.

One STAKEHOLDER is the Board of Directors. This particular stakeholder had a GOAL/CORE ISSUE: they want to increase the annual revenue. Using one or more ELICITATION TECHNIQUES, a REQUIREMENT to satisfy the above GOAL/CORE ISSUE was elicited: Predict how many customers will renew. One possible ELICITATION TECHNIQUE is a structured interview, but many others were possible (Fig. 2).

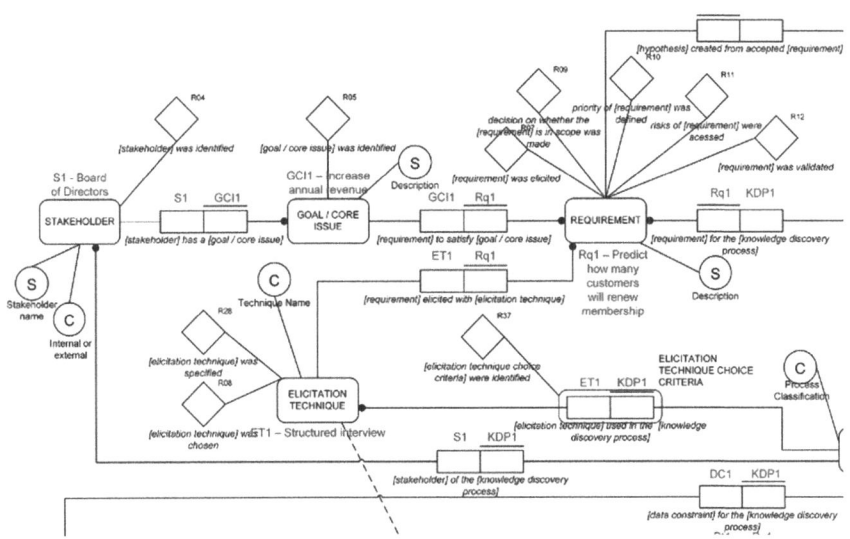

Fig. 2. Object Fact Diagram (Part 2)

Normally several STAKEHOLDERS will be identified (T04), each with one or more GOAL/CORE ISSUE from which several REQUIREMENTS will stem and be elicited (T05).

From the accepted REQUIREMENTS, we will then proceed to create a HYPOTHESIS that can be tested in a KDP. For this case, one of the tested HYPOTHESIS was if the number of logins can be used to predict if a customer will renew. The link between HYPOTHESIS and REQUIREMENTS is important for traceability.

In the end, the RESULT of the KDP will either confirm this hypothesis or not. For the KDP, there needs to be an estimation of COST AND RESOURCES, so that a Go-no-go decision (T14) can take place.

If the KDP proceeds, instances of classes corresponding to the DATA SOURCE (from which DATA will be selected and prepared), the data mining TOOL (in this case, Tableau 8.1), the type of DATA MINING TECHNIQUE (in this case, classification) and the ALGORITHM (in this case, AdaBoost) will be used to obtain a particular RESULT. The ALGORITHM might require a DATA MINING PARAMETER (or more) to be set. In this case we could change the value for a_t weight, but did not.

The KD AREA ARTEFACT is a generalization that includes ELICITATION TECHNIQUE, TOOL, DATA MINING TECHNIQUE, ALGORITHM and DATA MINING PARAMETER. The management of these artefacts (T27) involves specifying an artefact that was previously unknown to the system whenever needed (T28, T29, T30, T31, T32). The can then be chosen for use (T08, T20, T21, T22, T23) using ELICITATION TECHNIQUE CHOICE CRITERIA, TOOL CHOICE CRITERIA, DATA MINING TECHNIQUE CHOICE CRITERIA, ALGORITHM CHOICE CRITERIA or DATA MINING PARAMETER CHOICE CRITERIA respectively. It is important that the choice criteria are all documented, which is why all these classes appear (Fig. 3).

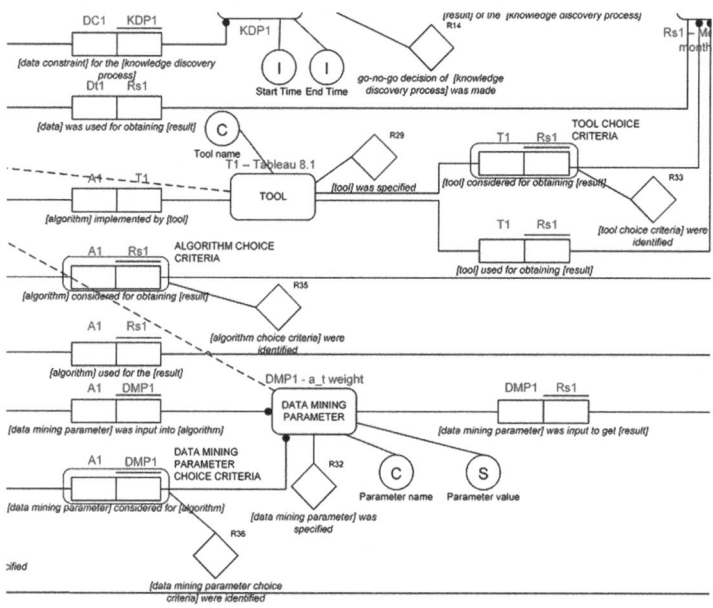

Fig. 3. Object Fact Diagram (Part 3)

From the DATA might result some kind of DATA CONSTRAINT. In this case, it was very noticeable that the customer age was not available. The identified DATA CONSTRAINTS affected the KDP.

As mentioned, the execution of a particular algorithm with particular parameters and applied to a particular data, in the context of a KDP will produce a particular RESULT

- for example, a classification model or a set of association rules. For the case study at hand, we found that the members who login more than once per month are more likely to renew.

The RESULT will be target of an analysis (T25). From such analysis the conclusion might be that new hypothesis needs to be formulated and/or new data, tools, data mining techniques or specific algorithms applied so that refined or alternative results are found. If none of this is necessary, the DEPLOYMENT of a RESULT can also be specified. For example, in this case it was decided that an annual report with the obtained result was to be produced (Fig. 4).

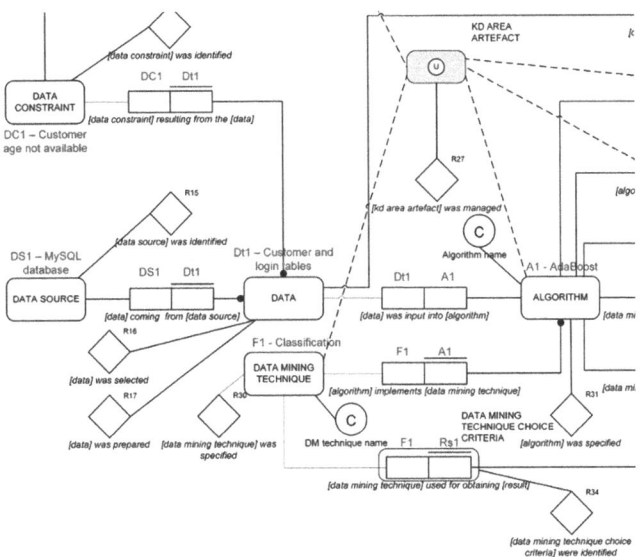

Fig. 4. Object Fact Diagram (Part 4)

3   Discussion and Conclusion

Other efforts have been made regarding knowledge discovery ontologies such as OntoDM [43] or Knowledge Discovery Ontology [44], but focus in great detail in the knowledge discovery process itself and don't show any particular insight regarding its surroundings, like the business side information.

This DEMO based ontology gives several interesting insights. Thanks to the specified classes, for a particular problem/opportunity we can keep a record of detailed and important information of a respective KDP. Namely keep a consistent and integrated record of important business side information like the stakeholders, requirements, hypothesis and costs; and also of the technical side like tools, sources and algorithms used. The class RESULT is pivotal in the sense that each instance will include not only the patterns obtained using the data mining technique, but also an analysis of the results which may lead to the formulation of new hypothesis and requirements on the business side.

Having SysPRE, an ontology that represents both the KD work in general and the RE for KD work specifically can help technical roles not lose track of the big picture while working on the task at hand. Also, since it is understandable not only by the technical roles involved, but also by other stakeholders, SysPRE can foster a more effective dialogue between them.

This ontology can encourage knowledge reuse of the KD process or RE KD process itself in a consistent and integrated fashion because it enables keeping a record of iterations and refinements of a particular process in a highly structured way. This way, it's hoped to make enterprises become aware of their own KD process and RE process in the KD projects, but also to improve such processes in reality, namely in terms of the success rate. In other words, this can help the lessons learned from the past be reused to improve the present.

The main contribution of this paper is to provide a systematization that can be applied to KD projects in general and to the requirements engineering process in such processes in particular.

Having a short, plain text description of a generic KD process with emphasis on RE that was proposed after doing a thorough literature review can be useful for novices in the area, both in the research and in the industry communities.

Having the SysPRE formal ontology can be helpful within the organization using them because it can:

- Enable keeping a record of iterations and refinements of a particular process in a highly structured way.
- Make enterprises (and specifically decision makers within the enterprise) become aware of their own KD process and RE process in the KD projects.
- Assist enterprises that want to improve their own KD process and RE process in the KD projects.
- Help each technical role involved keep an eye on the big picture while working on whatever task they are working on at that specific moment.

Having the SysPRE formal ontology can also be helpful for the communication between the organization and other stakeholders because, despite being formal, they are understandable and do sum up a lot of information in a graphical way.

Acknowledgments. This work was partially funded by FCT/MCTES LARSyS (UID/EEA/50009/2013 (2015-2017)).

References

1. The Standish Group, "1994 CHAOS Report," (1994)
2. Glass, R.L.: IT Failure Rates-70% or 10–15%? IEEE Softw. **22**(3), 110–112 (2005)
3. Jørgensen, M., Moløkken-Østvold, K.: How large are software cost overruns? A review of the 1994 CHAOS report. Inf. Softw. Technol. **48**(4), 297–301 (2006)

4. Glass, R.L.: The Standish report: does it really describe a software crisis? ACM Commun. **49**(8), 15–16 (2006)
5. Eveleens, J., Verhoef, C.: The Rise and fall of the Chaos report figures. IEEE Softw. **27**(1), 30–36 (2010)
6. Pohl, K.: Requirements Engineering: Fundamentals, Principles, and Techniques. Springer, Heidelberg (2010)
7. El Emam, K., Koru, A.G.: A replicated survey of IT software project failures. IEEE Softw. **25**(5), 84–90 (2008)
8. Atkins, C.: An Investigation of the Impact of Requirements Engineering Skills on Project Success. East Tennessee State University (2013)
9. Paiva, A., Varajão, J., Dominguez, C.: Principais aspectos na avaliação do sucesso de projectos de desenvolvimento de software. Há alguma relação com o que é considerado noutras indústrias? Interciencia **36**(3), 200–204 (2011)
10. Wateridge, J.: How can IS/IT projects be measured for success? Int. J. Proj. Manag. **16**(1), 59–63 (1998)
11. Fayyad, U.M., Piatetsky-Shapiro, G., Smyth, P.: Knowledge discovery and data mining: towards a unifying framework. KDD **96**, 82–88 (1996)
12. Royce, W.W.: Managing the development of large software systems. In: Proceedings of IEEE WESCON, vol. 26 (1970)
13. Statistics - YouTube. https://www.youtube.com/yt/press/statistics.html
14. Radicati, S. (ed.) Email Statistics Report 2013–2017 Executive Summary, April 2013
15. Manyika, J., Chui, M., Brown, B., Bughin, J.: Big Data: the Next Frontier for Innovation, Competition, and Productivity. McKinsey & Company, May 2011. http://www.mckinsey.com/insights/business_technology/big_data_the_next_frontier_for_innovation
16. Traverso, M.: Presto: interacting with petabytes of data at Facebook. Research at Facebook, November 2013. https://research.facebook.com/blog/1489667567986457/presto-interacting-with-petabytes-of-data-at-facebook/
17. Pytel, P., Britos, P., García-Martínez, R.: A proposal of effort estimation method for information mining projects oriented to SMEs. In: Poels, G. (ed.) CONFENIS 2012. LNBIP, vol. 139, pp. 58–74. Springer, Heidelberg (2013). doi:10.1007/978-3-642-36611-6_5
18. Inmon, W.H.: Building the Data Warehouse. Wiley, New York (2005)
19. Bernstein, A., Provost, F., Hill, S.: Toward intelligent assistance for a data mining process: an ontology-based approach for cost-sensitive classification. IEEE Trans. Knowl. Data Eng. **17**(4), 503–518 (2005)
20. Piatetsky-Shapiro, G.: Knowledge discovery in real databases: a report on the IJCAI-89 Workshop. AI Mag. **11**(4), 68 (1990)
21. Ganesh, M., Han, E.H., Kumar, V., Shekhar, S., Srivastava, J.: Visual Data Mining: Framework and Algorithm Development. Department of Civil Engineering, University of Minnesota, MN USA (1996)
22. Adriaans, P., Zantinge, D.: Data Mining. Addison-Wesley, Reading (1996)
23. Brachman, R.J., Anand, T.: Advances in knowledge discovery and data mining. In: Fayyad, U.M., Piatetsky-Shapiro, G., Smyth, P., Uthurusamy, R. (eds.) American Association for Artificial Intelligence, Menlo Park, pp. 37–57 (1996)
24. Berry, M.J., Linoff, G.: Data Mining Techniques: For Marketing, Sales, and Customer Support. Wiley, New York (1997)
25. Cabena, P., Hadjinian, P., Stadler, R., Verhees, J., Zanasi, A.: Discovering Data Mining: From Concept to Implementation. Prentice Hall, Upper Saddle River (1997)

26. Lee, S.W., Kerschberg, L.: A methodology and life cycle model for data mining and knowledge discovery in precision agriculture. In: IEEE International Conference on Systems, Man, and Cybernetics, vol. 3, pp. 2882–2887 (1998)
27. Buchner, A.G., Mulvenna, M.D., Anand, S.S., Hughes, J.G.: An internet-enabled knowledge discovery process. In: Proceedings of the 9th International Database Conference, Hong Kong, vol. 1999, pp. 13–27 (1999)
28. Wirth, R., Hipp, J.: CRISP-DM: towards a standard process model for data mining. In: Proceedings of the 4th International Conference on the Practical Applications of Knowledge Discovery and Data Mining, pp. 29–39 (2000)
29. SAS Institute: SEMMA (2005). http://www.sas.com/offices/europe/uk/technologies/analytics/datamining/miner/semma.html
30. Pyle, D.: Business Modeling and Data Mining. Morgan Kaufmann, San Mateo (2003)
31. Moyle, S., Jorge, A.: RAMSYS-A methodology for supporting rapid remote collaborative data mining projects. In: ECML/PKDD01 Workshop: Integrating Aspects of Data Mining, Decision Support and Meta-learning (IDDM-2001) (2001)
32. Solarte, J.: A proposed data mining methodology and its application to industrial engineering. Masters Theses, August 2002
33. Cios, K.J., Kurgan, L.A.: Trends in data mining and knowledge discovery. In: Pal, N.R., Jain, L. (eds.) Advanced Techniques in Knowledge Discovery and Data Mining, pp. 1–26. Springer, London (2005)
34. Gottgtroy, P.: Ontology driven knowledge discovery process: a proposal to integrate ontology engineering and KDD. (2007)
35. Rennolls, K., Al-Shawabkeh, A.: Formal structures for data mining, knowledge discovery and communication in a knowledge management environment. Intell. Data Anal. 12(2), 147–163 (2008)
36. Alnoukari, M., Alzoabi, Z., Hanna, S.: Applying adaptive software development (ASD) agile modeling on predictive data mining applications: ASD-DM Methodology. In: International Symposium on Information Technology, ITSim 2008, vol. 2, pp. 1–6 (2008)
37. Osei-Bryson, K.-M.: A context-aware data mining process model based framework for supporting evaluation of data mining results. Expert Syst. Appl. 39(1), 1156–1164 (2012)
38. IEEE Computer Society, "IEEE Standard Glossary of Software Engineering Terminology," IEEE Std 61012-1990, pp. 1–84, December 1990
39. Boehm, B.: A spiral model of software development and enhancement. SIGSOFT Softw. Eng. Notes 11(4), 14–24 (1986)
40. Martin, J.: Rapid Application Development. Mac Millan (1991)
41. IBM Rational software and systems delivery, 26 August 2014. http://www-01.ibm.com/software/rational/
42. Beck, K., Beedle, M., Bennekum, A.: Agile Manifesto (2001). http://www.agilemanifesto.org/
43. Panov, P., Soldatova, L., Džeroski, S.: OntoDM-KDD: Ontology for Representing the Knowledge Discovery Process. In: Fürnkranz, J., Hüllermeier, E., Higuchi, T. (eds.) DS 2013. LNCS (LNAI), vol. 8140, pp. 126–140. Springer, Heidelberg (2013). doi:10.1007/978-3-642-40897-7_9
44. Zakova, M., Kremen, P., Zelezny, F., Lavrac, N.: Automating knowledge discovery workflow composition through ontology-based planning. IEEE Trans. Autom. Sci. Eng. 8(2), 253–264 (2011)
45. Dietz J.L.: Enterprise ontology - understanding the essence of organizational operation. In: Chen CS., Filipe J., Seruca I., Cordeiro J. (eds) Enterprise Information Systems VII, pp. 19–30. Springer, Dordrecht (2007)
46. Piatetsky-Shapiro, G.: KDNuggets, "Poll: Data Mining Methodology," (2014). http://www.kdnuggets.com/polls/2014/analytics-data-mining-data-science-methodology.html

Revisiting the DEMO Transaction Pattern with the Unified Foundational Ontology (UFO)

Tanja Poletaeva[1(✉)], Giancarlo Guizzardi[2,3], João Paulo A. Almeida[3], and Habib Abdulrab[1]

[1] LITIS lab., INSA de Rouen, Rouen, France
ta.poletaeva@gmail.com, habib.abdulrab@insa-rouen.fr
[2] Free University of Bozen-Bolzano, Bozen-Bolzano, Italy
[3] Ontology and Conceptual Modeling Research Group (NEMO), Federal University of Espírito Santo, Vitória, ES, Brazil
{gguizzardi,jpalmeida}@inf.ufes.br

Abstract. In this paper, we revisit the DEMO transaction pattern in light of the domain-independent system of categories put forth by the Unified Foundational Ontology (UFO). In this process, we treat social relationships in the scope of the DEMO transactions as objectified social entities, and thereby separate the behavioural and structural aspects of the transaction pattern and clarify their interplay. Further, we represent the pattern in the OntoUML ontology-driven conceptual modeling language. The revisited pattern can be embedded in broader enterprise ontologies and reference conceptual models based in UML. The proposed OntoUML models can also be further refined to account for and consider different organizational implementations of business transactions. We demonstrate the proposed representation by applying it to OMGs EU-Rent case.

Keywords: Foundational ontology · Enterprise ontology · DEMO transaction pattern · Organizational implementation

1 Introduction

Since 1960s, conceptual modeling is widely adopted for knowledge communication among human users [1]. The importance of enterprise conceptual modeling in enterprise engineering and transformation [2] has encouraged the development of various enterprise modeling methods. Nowadays, there is a growing interest in approaches that employ ontologies as theoretical tools for improving conceptual models. Among such approaches, there is a mature DEMO methodology (the Design and Engineering Methodology for Organizations) [3], which comprises the DEMO enterprise ontology, the ontology-based enterprise modeling language, and the modeling method.

Despite the conceptual quality of DEMO, we observe that there are still opportunities for clarification and generalization of its conceptual basis, in particular considering some aspects of social relationships that evolve in business transactions. In addition to that, there are little guidelines on how to integrate knowledge conceptualized with DEMO to other (non-DEMO based) organizational conceptual models that are widely employed

© Springer International Publishing AG 2017
D. Aveiro et al. (Eds.): EEWC 2017, LNBIP 284, pp. 181–195, 2017.
DOI: 10.1007/978-3-319-57955-9_14

in practice (such as, e.g., reference organizational models captured in UML). These organizational conceptual models can play an important complementary role in the DEMO methodology, when used to represent the types of objects involved in business interactions and their properties in addition to elements of the DEMO conceptual models. Moreover, a broader ontological account of the DEMO models is required for modeling of organizational implementations in compliance with the core enterprise knowledge. We believe that coherent conceptual models of an organization at different levels of details support understanding and communication of desired organizational transformations.

We address these aforementioned issues in this paper. Firstly, we revisit a central notion of the DEMO enterprise ontology, namely, – the transaction pattern. The transaction pattern is a uniform communication pattern, which was proposed by Dietz for modeling of business interactions [3]. We revisit the nature of social relationships in the scope of the DEMO transaction pattern based on the domain-independent system of categories put forth by the Unified Foundational Ontology (UFO) [4]. In this process, we treat these social relationships as objectified social entities, and thereby separate the behavioural and structural aspects of the transaction pattern and clarify their interplay. Secondly, we represent the transaction pattern using the UFO-based OntoUML ontology-driven conceptual modeling language [4]. The revisited pattern can be embedded in broader enterprise ontologies and reference conceptual models based in UML. The proposed OntoUML models can also be further refined to account for and consider different organizational implementations of business transactions. We demonstrate the proposed representation by applying it to OMGs EU-Rent case that was the subject of DEMO analysis in [5].

OntoUML is an example of a conceptual modeling language whose meta-model has been designed to comply with the ontological distinctions and axiomatization of UFO [4, 6]. OntoUML has been successfully employed in a number of industrial projects in several different domains, such as petroleum and gas, digital journalism, complex digital media management, off-shore software engineering, telecommunications, retail product recommendation, and government [6]. A recent study shows that UFO is the second-most used foundational ontology in conceptual modeling and the one with the fastest adoption rate [7]. Moreover, the study also shows OntoUML is among the most used languages in ontology-driven conceptual modeling (together with UML, (E)ER, OWL and BPMN).

The outline of this article is as follows. In Sect. 2, we summarize the original DEMO ontological commitments related to the transaction pattern. In Sect. 3, we briefly explain some of the ontological foundations employed to revisit the pattern, including some key social notions in UFO, along with the fragment of OntoUML adopted here. In Sect. 4, we use the UFO notions to reconceptualize the DEMO transaction pattern with transactions analyzed from two complementary perspectives: a structural one, in which transactions are considered objectified social relations, and a behavioural one, in which transactions are considered occurrences or events. The section also presents OntoUML models of the transaction pattern, showing the interplay between two perspectives. Section 5 illustrates the application of the proposed OntoUML representation for modeling of organizational implementation variables, extending the models of Sect. 4. Finally, Sect. 6 presents our conclusions.

2   The DEMO Transaction Pattern

In the DEMO Enterprise Ontology [3], Dietz claimed to have proposed what he terms the "molecular structure" of business interactions. In his view, a business transaction is a minimal social conversation carried out between two social individuals, one of which (the *initiator*[1]) initiated the conversation in order to delegate an achievement of his/her goal to another party. The party who accepted the request for achievement of someone's goal is called the *executor*. In line with Habermas [8], Dietz relies on a desire of both parties to reach a consensus in a business deal.

Hereafter, we summarize the ontological commitments of the DEMO Enterprise Ontology [3, 9] related to the transaction pattern.

- C1. By performing *coordination acts* (*C-acts*), social individuals of an enterprise enter into and comply with social commitments towards each other regarding the product to be brought about.
- C2. The original new thing that is created by a C-act is a commitment (also named a *coordination fact*).
- C3. Two time aspects of coordination facts are distinguished: the event time and the settlement time.
- C4. By performing *production acts* (*P-acts*), social individuals in an organization create products.
- C5. There is a one-to-one relationship between *transaction kinds* and product kinds. Transaction kind is a basic property of every *transaction*.
- C6. *Actor role* is the authority to fill the *executor role* in transactions of a particular transaction kind. It includes (by definition) the authority to be the initiator in transactions of a number of (other) transaction kinds.
- C7. An *actor* is a social individual (subject) in the quality of filling an actor role.
- C8. A transaction involves two actors, one as the *initiator* and one as the *executor*.
- C9. The *process of a transaction* is a temporally ordered sequence of coordination acts of the initiator and the executor, starting from a requesting coordination act of the initiator.
- C10. The process of a transaction is a path, possibly including iterations, through a *universal transaction pattern*.
- C11. A complete transaction goes off in three consecutive phases: the *order phase*, the *execution phase*, and the *result phase*. The process of a transaction in the order and the result phases is a sequence of coordination acts. In the execution phase, the executor performs some production act(s).

---

[1] In this section, we introduce the terms from the DEMO vocabulary in *italics*.

3   UFO and OntoUML

In this section, we present a subset of OntoUML language that is employed here for the representation of the DEMO transaction pattern. We also briefly discuss the UFO concepts underlying the OntoUML constructs used. Finally, we summarize the UFO ontological commitments about social entities [10] that are relevant to the DEMO commitments discussed in Sect. 2.

3.1   OntoUML

OntoUML is an ontologically well-founded version of UML (Unified Modeling Language) whose metamodel reflects a number of ontological distinctions and axioms put forth by UFO [4, 6]. This means that an OntoUML representation of state of affairs in reality is unambiguously interpreted based on domain-independent ontological categories.

In OntoUML, class constructs stereotyped by «Kind» represent object types that supply a uniform *principle of identity*[2] for their instances. Specializations of classes representing kinds are stereotyped as «SubKind», «Role», or «Phase». All these specializations inherit their principle of identity from «Kind» types. While object types stereotyped by «Kind» and «SubKind» are necessarily applied to their instances in every possible world (i.e., these types are *rigid*), instances of «Role» and «Phase» types can cease to be instances of these types without ceasing to exist and without altering their identity. Moreover, while instances of «Phase» types are characterized by a change of their intrinsic property(s), instances of «Role» types are characterized by a relational property(s) acquired in relationships with other entities.

«Category» and «RoleMixin» types represent an abstraction of properties that are common to multiple «Kind» types and, therefore, do not carry a unique principle of identity for their instances. Properties associated with «Category» types are rigid and relationally-independent, while properties associated with «RoleMixin» types necessarily represent an abstraction of contingent (or *anti-rigid*) properties that are common to different «Role» types.

In addition to the aforementioned object types, OntoUML class elements represent types of existentially dependent individuals that can only exist by inhering in other individuals. Such individuals are called *moments*. Those moments that inhere in one single individual are categorized as «Mode» or «Quality» types. While qualities (also called *individual qualities*) are moments that change in a particular space of possible values (e.g. a color, a temperature, a weight), modes are complex individual moments that can have their own qualities that take their respective values in multiple independent value dimensions (e.g., a symptom, a capacity, a complex intention). While inhering in a single individual, some modes and qualities can *externally depend on* (possibly a multitude of) other individuals that are independent from their bearers. Moments that existentially depend on two or more individuals are categorized as «Relator» types.

---

[2] The terms from the UFO vocabulary, which are introduced in addition to OntoUML stereotypes, are highlighted in *italics*.

Instances of «Event» types are *perdurants*. Perdurants unfold in time accumulating temporal parts. They are defined by the sum of their parts (their constituent sub-events) and they bear to each other a number of temporal ordering and causality relations. Moreover, perdurants are *manifestations* of dispositional properties of moments (qualities, modes, and relators). Finally, perdurants are immutable in all their parts and all their properties [11].

Moments are connected to their bearers via existential dependence relations. In OntoUML, intrinsic moment types (quality and mode types) are connected to the type representing their bearers via a relation of «characterization». This relation is mapped at the instance level to a relation of *inherence*, i.e., a particular type of functional external dependence relation; relator types, in contrast, are connected to the type of entities they relate (bind, mediate) via an association stereotyped as «mediation», which is mapped at the instance level to a non-functional type of existential dependence relation.

3.2 UFO-C: The Social Layer of UFO

In a social context, the UFO-C part [10] of UFO distinguishes between agentive and non-agentive substantial individuals. Agentive individuals (or *agents*) are capable of bearing special kind of moments named *intentional moments*. Intentional moments can be further specialized into *mental moments* (including *beliefs*, *desires* and *intentions*) and *social moments*. Moreover, each type of intentional moments necessarily has a propositional content, which may be matched by certain situations in reality. Thus, the intentionality of agents should be understood as the capacity of their properties to refer to possible situations of reality.

Among other types of intentional moments, *Intentions* refer to desired state of affairs to which an agent *internally commits* at pursuing. For this reason, intentions cause the agent to perform *actions*. Actions are intentional events, i.e., events with the specific purpose of satisfying the propositional content of some intention of an agent. The propositional content of an intention is termed a *goal*. UFO contemplates a relation between situations and goals such that a situation may satisfy a goal.

Communicative acts (special kinds of actions) can create *social moments* (*commitments* and *claims*) inhering in the agents involved in these communicative acts. In opposite to internal intentional moments, social commitments and claims *inhere* in one agent and are *existentially depend* on another. If a social commitment inheres in an individual X and is externally dependent on another agent Y (i.e., it is a commitment of X towards Y) then there is a dual *social claim* inhering in Y and which is externally dependent of X (i.e., the claim of Y towards X). In other words, commitments and claims always form a pair that shares a unique propositional content [10]. Two or more pairs of mutually dependent commitments and claims form a kind of social relationship between involved social individuals. This social relationship is termed in UFO-C a *social relator*. Social commitments and claims are often associated with internal commitments (self-commitments).

4   Ontological Analysis and Representation of the Transaction Pattern

In this section, we propose to align the original ontological commitments for the DEMO transaction pattern given in Sect. 2 employing the conceptual notions put forth by UFO. In this process, we elaborate on the benefits of considering business transactions as endurants (more specifically as social relators) together with reifying corresponding transaction events as the context of business interactions. In addition to revisiting the conceptual aspects of the transaction pattern in light of the UFO, we provide a representation of the revisited pattern using OntoUML models which can be later used as a basis for extension in order to model enterprise-specific settings. In this section as well as in Sect. 5, we write the elements of the proposed models in *italics*. Stereotypes and the names of relations start with lowercase, while types are capitalized.

4.1   Transactions as Endurants

A central notion in the transaction pattern is the notion of transaction. We propose a transaction should be understood as a relator, composed of social commitments and claims made by involved actors in their negotiation about an achievement of some shared goal (i.e., a production result), as well as other relational qualities acquired by actors in this negotiation. Thus, a transaction can be represented by a relator mediating two actors, which play the roles of *Initiator* and *Executor* (Fig. 1). A particular role played by an agent in a transaction is defined by the type of his commitments and claims. An actor is the executor, when he commits himself to a requesting actor (the initiator) to achieve a production result. Although in this paper we excluded the self-activating actors from consideration, this additional constraint cannot be expressed in UML.

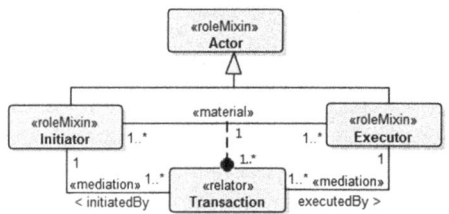

Fig. 1. Participation in a transaction

Actor specializes the notion of UFO Agent; an *Actor* is an Agent that participates (contingently) in a *Transaction*, which suggests that the *Actor* type should be stereotyped «role». We have opted to represent the type *Actor* as a *role mixin* (instead of *role*) in order to cater for the possibility that they obey different principles of identity (for example to allow for individuals and for teams to be considered *Actors*).

By applying the powertype pattern from a Multi-Level Theory [12] to C6, C7, and C8 commitments from Sect. 2, we propose modeling of the initiator and the executor of a transaction as instances of the *Actor Role* powertype (Fig. 2), i.e. a rigid sortal whose instances are types. Following [13], we extended the OntoUML metamodel by introducing the stereotype «hou» to represent high-order universals. All roles specializing *Initiator* are instances of the *Initiator Actor Role* powertype, while all roles specializing *Executor* are instances of the *Executor Actor Role* powertype. The generalization set of *Initiator Actor Role* and *Executor Actor Role* is overlapping, which means that some instances of *Transaction Kind* relate with only one instance of the *Actor Role* powertype.

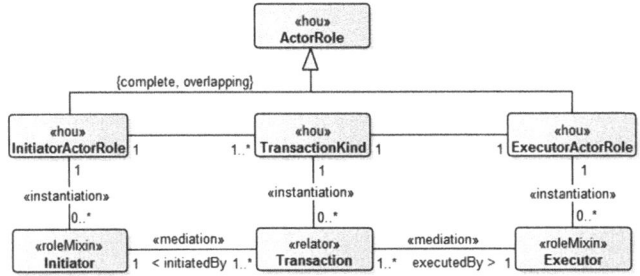

Fig. 2. Initiator and executor actor roles in relation to transaction kind

Following DEMO, we assume that an actor playing a particular actor role in an organization commits himself to perform coordination acts and to accept social commitments of certain kinds under certain types of situations. Hereafter, social commitments resulting coordination acts are called *C-commitments* (C- for "coordination"). Moreover, since social commitments and claims always appear in pairs, we refer to *C-claims* that result from coordination acts in addition to C-commitments. Note that we do not use the term "C-fact" originally proposed in DEMO as the result of coordination acts. This substitution of terms is motivated by the considerable difference in understanding of the notion of fact put forth by the UFO and the DEMO ontological commitments.

The model depicted in Fig. 3 facilitates explicit representation of C-act types, C-commitment and C-claim types which constitute a transaction. The details about relations of endurants and events can be found in [14].

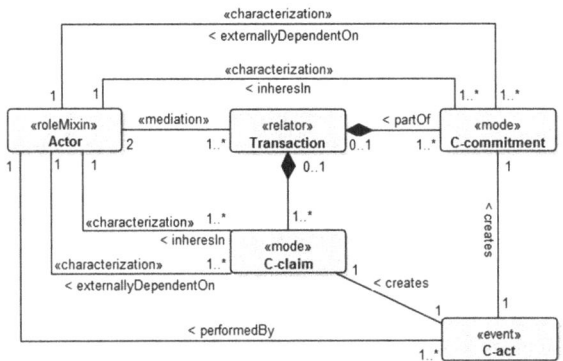

Fig. 3. Considering transactions as endurants composed of C-commitments and C-claims

Subsequent coordination acts performed by actors involved in a transaction contribute to the life of this transaction over time, i.e., to the changes it might undergo. Reified transaction types in conceptual models allow explicit representation of transaction phases and properties. A transaction phase is defined by C-commitments that constitute this transaction during a certain period of time. For instance, a transaction can be suspended until a C-commitment of a particular type is created, or it can be terminated, etc. As it is the case for all endurant types, transaction phases are represented on conceptual models as specializations of a transaction kind.

By considering transactions as endurants, we are able to specify their qualities in addition to qualities of the participating actors. For example, a yearly membership registration (*Transaction*) of a customer (*Initiator*) in a company (*Executor*) can have a particular cost (a quality inhering in the transaction itself). Further, a modeler can characterize changes of a transaction cost over time. Another example is characterization of transactions by a status that can be, for instance, "successful", "failed", or "unconfirmed".

As full-fledged endurants, transactions can play roles [15]. For instance, outsourced transactions acquire specific contingent properties being parts of organizational structures of third-party companies. Transactions initiated externally (i.e., by the initiator from the environment of the organization under consideration) can play a role of service agreements [16] provided by the organization.

4.2 Transactions as Events

According to the C9 and C10 DEMO commitments, a transaction process extends in time by accumulating its temporal parts similar to other perdurants. In [11], it is proposed to consider perdurants (there, generally termed events) as the manifestation of individual qualities and relationships. Taking into account this notion of events, a transaction process is a *Transaction Event* focusing on relationships of actors involved in a minimal business interaction (i.e., on a transaction). In this section, we elaborate on the practical relevance of having a transaction event as a modeling construct. Our arguments are based on those in [11] made in favour of reifying events as the context of relationships.

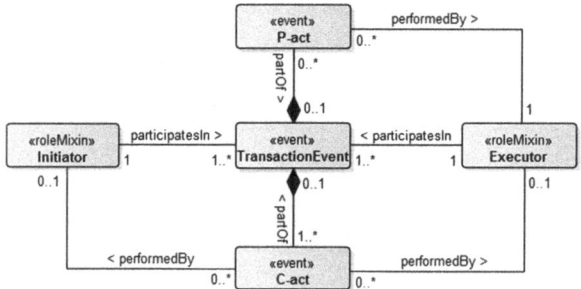

Fig. 4. Considering transactions as events composed of C- and P-acts

Hereafter, we consider a transaction event constituted by coordination acts of involved actors. These C-acts result in coordination commitments and claims, which, in turn, constitute the transaction in the focus of the transaction event. Taking into account that "roles are usually understood as ways of participation[3] to an event" [11], the C7 commitment can be reformulated as follows: *an actor is a social individual which participates a transaction event by making C-acts* (Fig. 4).

The UML composition relation (represented by an association having a black diamond in the association end connected to the class representing the whole) in Fig. 4 implies that the parts in the depicted part-whole relations are non-shareable among things of that whole class, i.e., that the maximum cardinality w.r.t. to the whole class is 1. A transaction event should be understood as an optional whole for coordination acts.

The execution phase of a transaction (see Sect. 2) is constituted by a production act, and it can be modeled as a proper part of a transaction event. Considering the execution phase as an event allows unambiguous relation of this phase with the completion (events) of other transactions.

A modeler may want to explicitly represent other temporal constraints concerning a transaction process. For instance, one may want to introduce a pick up event as a proper part of a delivery transaction in order to express the constraints of the transaction duration. When a transaction event evolves in time, it goes through phases composed of events. Contrary to transaction phases composed of C-commitments and C-claims (i.e., endurants), these phases are complex events. By explicitly represented events (i.e., transaction events, C-acts, and P-acts), we can represent temporal and causal relations between them [14].

Finally, the consideration of a transaction event in a broader context of the involved objects and their qualities facilitate the specification of constraints for constituting coordination acts. For instance, a commitment for a delivery can be restricted by values of a requested drop-off location of a delivered product.

5   Applying the Transaction Pattern: A Case Study

We believe that the conceptual models and ontological distinctions proposed in previous sections facilitate understanding and communication of various organizational implementations. In [5], the authors provided a thorough analysis of organization implementation descriptions of OMGs EU-Rent case [17], using the DEMO construction model of a fictitious car rental company for the representation of implementation independent organizational essence. In this section, (a part of) the analysis given in [5] is supplemented by OntoUML representations, which were obtained by specializing the modeling structures from in Sect. 4. Since we did not transform the DEMO modeling language to OntoUML, the conceptual models in this section should be considered as additions to the model depicted in Fig. 5.

---

[3] Following [11], we understand participation as a formal relation linking endurants and perdurants.

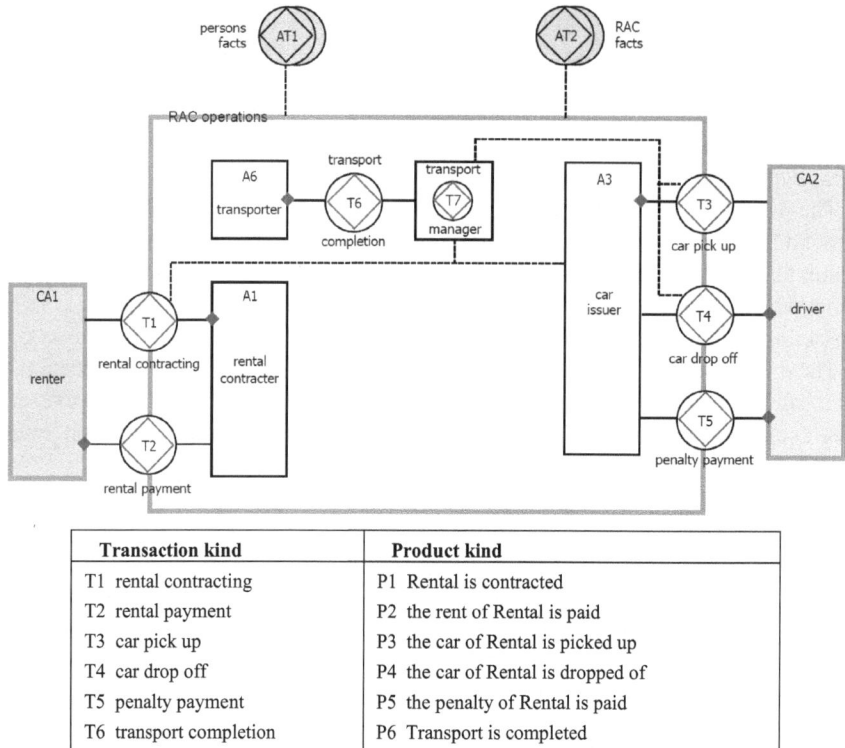

Transaction kind	**Product kind**
T1 rental contracting	P1 Rental is contracted
T2 rental payment	P2 the rent of Rental is paid
T3 car pick up	P3 the car of Rental is picked up
T4 car drop off	P4 the car of Rental is dropped of
T5 penalty payment	P5 the penalty of Rental is paid
T6 transport completion	P6 Transport is completed
T7 transport management	P7 transport management for Day is done

Fig. 5. The organization construction diagram (OCD) and the Transaction Result Table (TRT) of a car rental company (after [5])

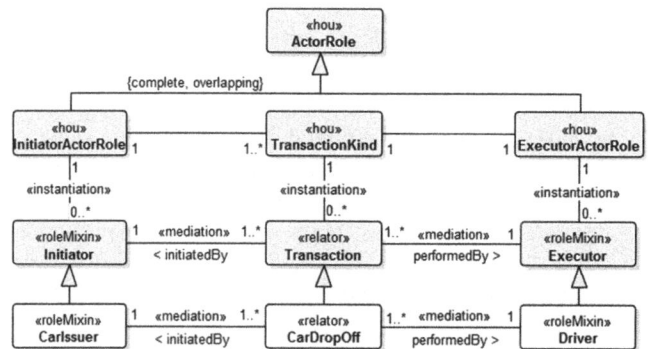

Fig. 6. *Car Drop Off* transactions as objectified social relationships

The DEMO construction model (Fig. 5) representing the immutable ontological essence of a fictitious car rental company "Rent-A-Car" (of RAC for short) was borrowed from [5]. Hereafter, we elaborate on some organization implementation choices with the references to this ontological essence. A reading guide for this model can be found in [18].

According to the DEMO analysis, "car drop off" is one of the transaction kinds in RAC, of which "car issuer" and "driver" are the participating actor roles (see Fig. 5). By specializing the model depicted in Fig. 2, a modeler can explicitly express that "car drop off" is an instance of *Transaction Kind*, and instances of the *Car Drop Off* type are transactions (Fig. 6); actors playing the *Car Issuer* role initiate *Car Drop Off* transactions, while actors playing the *Driver* role execute them. In Fig. 6 and other figures of this section, we highlight the elements of the models from Sect. 4 in grey.

Case description in [5] provides further details about *Car Drop Off* transactions. The RAC company allows cars to be dropped-off in different locations. Students are hired to implement this service: *"For a small amount of money, a student would await the arrival of a rented car, e.g. at an airport, and drive it back to the office of RAC, after which the student would go home by public transport"* [5].

The given implementation of the company was analyzed in [5] as follows. *"... Students are authorized to accept the drop-off, so there is an assignment between employees and act types (during some time frame), and, as the student is not the requester of the drop-off, there is some form of delegation..."* [5]. Some organization implementation variables were extracted from this description including V1[4] and V2:

- V1: Cross-reference which employee is allowed to perform which type of act (cross reference functionary type/act type);
- V2: Delegation of act types.

In order to model the values of these variables actual for RAC, we explicitly represent a type of C-commitments (*Accept-commitment*), which are created when drop-offs are accepted in the scope of *Car Drop Off* transactions (Fig. 7). In this specification, the acceptance of a drop-off may not happen (in case the transaction has been failed at some moment). Although instances of *Car Issuer* are recognized as bearers of C-commitments of the *Accept-commitment* type, the execution of related C-acts can be delegated to students. As illustrated in Fig. 8, *Accept-act* C-acts are performed by instances of *Student Qua Car Issuer*, which participate in *Car Drop Off Event* transaction events.

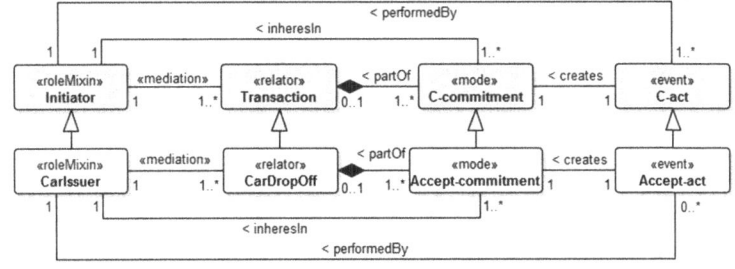

Fig. 7. *Car Issuer* participations in *Car Drop Off* transactions

---

[4] Here, we do not support the original numbering of implementation variables from [5].

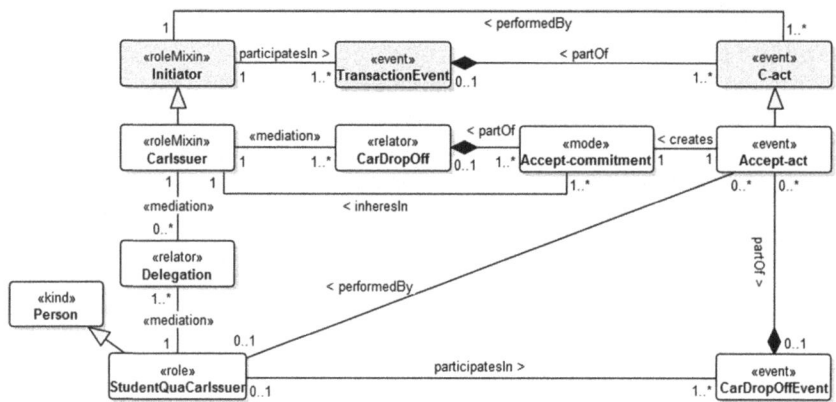

Fig. 8. Accepting coordination acts of *Car Drop Off* transactions delegated to students

Without further elaboration on a relation between actor roles and organizational roles (like employee, student, or director), we specify the delegation relationship between *Car Issuer* and the *Student Qua Car Issuer* organizational role, where the latter can be thought of as a specialization of the *Student* type. Contrary to *Car Issuer*, instances of the *Student Qua Car Issuer* type are identifiable persons. This fact is represented by the *role* stereotype and the kind *Person*, which is a supertype of this role. Despite participating in car drop off events (instances of the *Car Drop Off Event* type), instances of *Student Qua Car Issuer* do not bear social commitments created by accepting coordination acts.

As noted in [5], *"…the drop-off location could be anywhere (airport departure hall 3, town center, …) and not necessarily a RAC office. This implies that the state and accept of the drop-off can happen at any location. For that, the locations of performing certain acts must be defined."* Another organization implementation variable was defined accordingly as follows:

– V3: Cross-reference which act type can be performed in which location (event location restrictions).

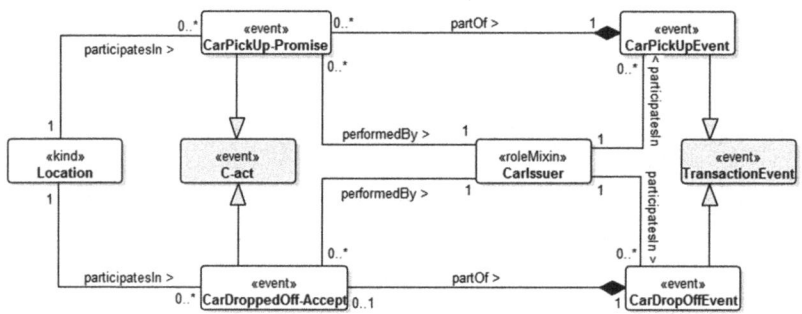

Fig. 9. Location constraints of C-acts

The behavioural aspects of the transaction pattern can be referred to for grounding the discussions on values of this implementation variable. Obtained by specializing the model in Fig. 4, the lower half of the model in Fig. 9 expresses a constraint concerning the location where the acceptance of a dropped car (an instance of *Car Dropped Off-Accept*) can be performed by an actor playing the *Car Issuer* actor role. The location constraint cannot be applied to the *Car Drop Off* type (see Fig. 5), since a location is not directly involved in transactions [11].

The upper and the lower parts of the model in Fig. 9 together illustrate the interdependency of the location constraints of C-acts constituting transaction events of different types. Based on this model, an enterprise modeler can further specify to which extent the execution of C-acts of a particular type is restricted to the location at which the whole transaction event is triggered. This interdependency can further be specified, e.g., "*pick-up can only be done at branches near airports, while drop-off can be done at any branch*" [5]. For the given implementation of RAC, the *CarPickUp-Promise* event can be constrained by *Airport Area*, which is a specialization of *Location*.

One can imagine the implementation of RAC, in which the acceptance of a car by a student playing the *Student Qua Car Issuer* actor role triggers the planning of a maintenance control required from a mechanical engineer, i.e., a *Transport Completion* transaction (see Fig. 5).

The notion of transaction event from Sect. 4 allows referring to a transport completion transaction before a transport manager has initiated it. In its planning stage, a transport completion transaction can be represented by a transport completion transaction event (Fig. 10).

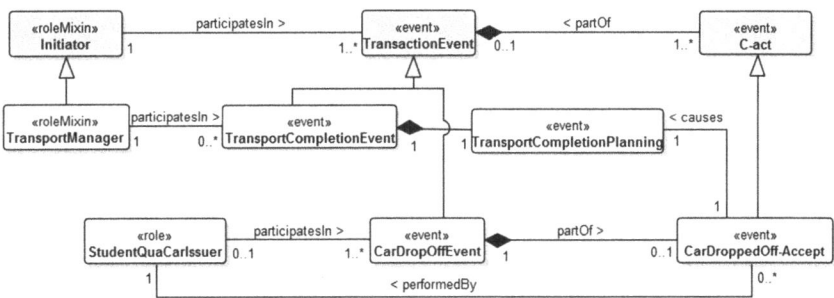

Fig. 10. Interrelation of *Car Drop Off* and *Transport Completion* transactions

6   Final Considerations

Enterprise conceptual modeling is not an easy task. In this paper, we make an attempt to prepare foundations for possible extensions of the DEMO-based conceptual models by considering the OntoUML representation of the DEMO transaction pattern. By using the proposed modeling constructs, we represent some implementation aspects of a fictitious Rent-A-Car company and demonstrate the ability of these constructs to forth making important ontological distinctions that can be overlooked in ordinary textual descriptions of implementations. We also demonstrate the benefits of the incorporation of events into structural conceptual models.

Acknowledgments. This research is partly funded by the FEDER grant number HN0002134: CLASSE 2 ("Les Corridors Logistiques: Application a la Vallee de la Seine et son Environnement"). This research is also partly funded by the Brazilian Research Funding Agencies CNPq (grants number 311313/2014-0 and 461777/2014-2) and FAPES (grant number 69382549).

References

1. Wand, Y., Weber, R.: Research commentary: information systems and conceptual modeling – a research agenda. Inf. Syst. Res. **13**(4), 363–376 (2002). INFORMS, USA
2. Barjis, J.: Enterprise modeling and simulation within enterprise engineering. J. Enterpr. Transform. **1**(3), 185–207 (2011). Taylor&Francis Online
3. Dietz, J.L.G.: Enterprise Ontology – Theory and Methodology. Springer, Heidelberg (2006)
4. Guizzardi, G.: Ontological foundations for structural conceptual models. Telematics Instituut Fundamental Research Series, ISSN 1388-1795, No. 015, The Netherlands (2005)
5. Land, M., Krouwel, M.: Exploring organizational implementation fundamentals. In: Proper, H.A., Aveiro, D., Gaaloul, K. (eds.) EEWC 2013. LNBIP, vol. 146, pp. 28–42. Springer, Heidelberg (2013). doi:10.1007/978-3-642-38117-1_3
6. Guizzardi, G., Wagner, G., Almeida, J.P.A., Guizzardi, R.S.S.: Towards ontological foundation for conceptual modeling: the unified foundational ontology (UFO) story. Appl. Ontol. **10**(3–4), 259–271 (2015). IOS Press
7. Verdonck, M., Gailly, F.: Insights on the use and application of ontology and conceptual modeling languages in ontology-driven conceptual modeling. In: Comyn-Wattiau, I., Tanaka, K., Song, I.-Y., Yamamoto, S., Saeki, M. (eds.) ER 2016. LNCS, vol. 9974, pp. 83–97. Springer, Cham (2016). doi:10.1007/978-3-319-46397-1_7
8. Habermas, J.: The Theory of Communicative Action. Lifeworld and System: A Critique of Functionalist Reason, vol. 2. (Translated by Thomas McCarthy), 3d corrected edn. 1985. Suhrkamp Verlag, Frankfurt am Main (1981)
9. Dietz, J.L.G.: The PSI theory – understanding human collaboration. Technical report TR-FIT-15-05. Faculty of Information Technology Czech Technical University in Prague (2015). http://www.ciaonetwork.org/uploads/eewc2015/ee_theories/theories/. Accessed 2016
10. Guizzardi, G., Falbo, R.A., Guizzardi, R.S.S.: Grounding software domain ontologies in the unified foundational ontology (UFO): the case of the ODE software process ontology. In: XI Iberoamerican Workshop on Requirements Engineering and Software Environments, pp. 244–251 (2008)
11. Guarino, N., Guizzardi, G.: Relationships and events: towards a general theory of reification and truthmaking. In: Adorni, G., Cagnoni, S., Gori, M., Maratea, M. (eds.) AI*IA 2016. LNCS (LNAI), vol. 10037, pp. 237–249. Springer, Cham (2016). doi: 10.1007/978-3-319-49130-1_18
12. Carvalho, V.A., Almeida, J.P.A., Guizzardi, G.: Using a well-founded multi-level theory to support the analysis and representation of the powertype pattern in conceptual modeling. In: Nurcan, S., Soffer, P., Bajec, M., Eder, J. (eds.) CAiSE 2016. LNCS, vol. 9694, pp. 309–324. Springer, Cham (2016). doi:10.1007/978-3-319-39696-5_19
13. Falbo, R.A., Ruy, F.B., Guizzardi, G., Barcellos, M.P., Almeida, J.P.A.: Towards an enterprise ontology pattern language. In: 29th Annual ACM Symposium on Applied Computing, pp. 323–330. ACM (2014)

14. Guizzardi, G., Wagner, G., Almeida Falbo, R., Guizzardi, R.S.S., Almeida, J.P.A.: Towards ontological foundations for the conceptual modeling of events. In: Ng, W., Storey, V.C., Trujillo, J.C. (eds.) ER 2013. LNCS, vol. 8217, pp. 327–341. Springer, Heidelberg (2013). doi:10.1007/978-3-642-41924-9_27

15. Guarino, N., Guizzardi, G.: "We need to discuss the *relationship*": revisiting relationships as modeling constructs. In: Zdravkovic, J., Kirikova, M., Johannesson, P. (eds.) CAiSE 2015. LNCS, vol. 9097, pp. 279–294. Springer, Cham (2015). doi:10.1007/978-3-319-19069-3_18

16. Nardi, J.C., De Almeida Falbo, R., Almeida, J.P.A., Guizzardi, G., Ferreira Pires, L., Van Sinderen, M.J., Guarino, N., Fonseca, C.M.: A commitment-based reference ontology for services. Inf. Syst. **54**, 263–288 (2015). Elsevier Ltd.

17. Object Management Group: Business Motivation Model (BMM) Specification, V1.1. OMG Available Specification OMG Document Number: formal/2010-05-01 (May 2010). http://www.omg.org/spec/BMM/1.1/PDF/

18. Op't Land, M., Dietz, J.L.G.: Benefits of enterprise ontology in governing complex enterprise transformations. In: Albani, A., Aveiro, D., Barjis, J. (eds.) EEWC 2012. LNBIP, vol. 110, pp. 77–92. Springer, Heidelberg (2012). doi:10.1007/978-3-642-29903-2_6

Organisation Design

An OD-Pearl for the EE-Oyster

L.J. Lekkerkerk(✉)

Institute for Management Research, Radboud University, Nijmegen, The Netherlands
h.lekkerkerk@fm.ru.nl

Abstract. Explaining the basics of Lowlands-SocioTechnical Systems Design (L-STSD) to the Enterprise Engineering community is the first goal of this contribution in order to show how it fits with EE. Then, a first attempt is made to link L-STSD to basic EE insights gained so far. It seems that the commonalities are good, because of a shared basis in general systems theory, and a desire to optimize organizational functioning. The differences may not be problematic, so L-STSD may be a compliant view to enhance the EE-body of knowledge with a new lens on organisational structure, or the labour to be divided and coordinated & controlled.

Keywords: Organisation design · Organisational structure design · Sociotechnical systems design

1 Introduction

At the 7th Enterprise Engineering Working Conference 2017 a special session was planned on Organizational Design, with the intent to explore the OD-field for opportunities to enrich the Enterprise Engineering (EE) body of knowledge. The author is especially familiar with the Lowlands SocioTechnical Systems Design (L-STSD) approach to Organisational Design and is a novice in the EE-field. Table 1 presents EE-phrases with their equivalent from L-STSD sources. The similarities will be obvious to the reader.

Hence, explaining the basics of Lowlands-SocioTechnical Systems Design (L-STSD) to the EE-community is the first goal of this paper. Because L-STSD is little known outside Dutch speaking countries, The Netherlands and Flanders, due to a lack of publications in English, this requires a textbook-like summary, which is, admittedly, somewhat unusual for a conference paper. This summary then serves as a starting point for the second goal, comparing and contrasting L-STSD with EE to figure out whether and how EE may be enriched by L-STSD (and vice versa). This paper only presents a first brief comparison by the author based on his first novice understanding of EE. At the EEWC-2017 the debate will start, and hopefully lead to further developments.

Organisations seem to pay more attention to structure. Since around 2010 managers try to improve their organization's structure using one of the hypes, e.g. holacracy, agile/scrum, 'Spotify', teal organizing [16], or Lean, hoping to meet the higher challenges of contemporary multiple value creation by introducing forms of self-managing teams, and

© Springer International Publishing AG 2017
D. Aveiro et al. (Eds.): EEWC 2017, LNBIP 284, pp. 199–219, 2017.
DOI: 10.1007/978-3-319-57955-9_15

flattening the hierarchy. However, to enable the successful introduction of self-manage-ment and a flat structure the organizational structure needs to be redesigned in most cases, but the approaches mentioned tend to overlook that. Also, not all managers should design an organisational structure. Everybody will agree that designing an affordable, safe and durable bicycle, car, software package, or computer requires a design team with specific knowledge, gained from a four to five year engineering study, and much expe-rience. Most managers do not have an education as organizational structure engineer. And when moving the boxes in the organization chart, merely changing reporting lines, those managers may really believe that they are 'redesigning the structure' of their organization. As this paper shows, there is more to organisation design than 'redrawing the chart', and this lack of knowledge will in part explain why so many reorganizations fail miserably.

Table 1. EE and L-STSD have similar views on organisations

	EE views	L-STSD-views
1	'A holistic and general system theory based understanding of how to (re)design and run enterprises effectively'	General systems theory based, and favours holistic redesign of the whole organization, rather than just redesigning a lower level unit
2	Referring to organizations as socio-technical systems	Named in Dutch 'moderne sociotechniek', in English we refer to it now as 'Lowlands sociotechnical systems design' or L-STSD
3	The human beings that are the 'pearls' of the organization	Aiming at improving the 'quality of work' so that the human employees have better jobs
4	Acknowledging the fact that these pearls do not always act purely rational or evidence-based	Eliminating unnecessary 'red tape' and empowering employees to contribute to the purpose of the organisation, and improving the 'quality of work'
5	A desire to make change efforts more successful, and also an enterprise as a whole more effective and efficient	Using a design sequence and design rules that lead to higher controllability (both operational and strategical), contributing to a more effective and efficient organisation
6	Welcoming proposals to include compliant views into the body of knowledge	N.A.
	[6, 7, 13, 14, 20]	[1, 10–12, 15, 17–19, 22, 24, 25]

2   Backgrounds and Basics of Lowlands SocioTechnical Systems Design

2.1   Other OD-Approaches and Why L-STSD

There are different approaches to Organization Design, or rather Organizational Struc-ture Design, that may be used for an OD-EE-comparison. After mentioning a few OD-approaches, L-STSD is put forward as the most suitable candidate for this goal.

The configurational approach describes several archetypes of structures, like the five of Mintzberg [21], that each matches a typical situation, but the idea that 'five sizes fit all' is flawed. Another one is the contingency approach in which the structure design should match the internal and external contingencies of the organization. Burton et al. [5] is a typical example of this approach. Contingencies are different for each organisation, but some patterns emerged from studying structures. Then there is Galbraith's [9] information processing view, which sees the organizational structure mainly as a tool that must be designed in such a way that it can process the necessary amount of information processing to match the information need of the different organizational units. Next a process oriented approach like Business Process Redesign mainly looks at organizations that are information processors, and tries to make units with an end-to-end responsibility for a value adding process, instead of a structure in which units perform similar activities (activity based structure also labelled functional or bureaucratic structure). Lean Thinking, originating in the automobile industry, also has a (primary) process-focus and also tries to form units of people and equipment dedicated to and responsible for a 'flow', delivering a subset of high customer value products.

Finally the approach chosen for this paper, is called Lowlands SocioTechnical Systems Design (L-STSD). Founding father Ulbo De Sitter and his co-workers hardly published in English. It is rooted in the sociotechnical tradition started by Tavistock Institute research by Bamfort, Emery, and Trist, in the English coal mines in the 1950's. Ulbo de Sitter further developed sociotechnical thinking in the 1980's and 1990's into an integral organisational structure design approach, with a theoretical foundation in systems theory and organizational cybernetics (e.g. Ashby [2]). It is aimed at improving the controllability of the organization and in doing so improving both Quality of Work and Quality of the Organization, which comes close to the idea of 'joint optimization' of the social and technical subsystems the early sociotechnical writers were advocating. De Sitter's Dutch version of sociotechnical systems design (D-STSD), was sometimes named 'integral organizational renewal' (IOR, e.g. in de Sitter et al. [25]). Following jargon used in the US-based but worldwide SocioTechnical Systems Roundtable, 'sociotechnical systems design' was adopted by Van Amelsvoort (in Mohr and Van Amelsvoort ed's 2016 [22]), and Lowlands replaced Dutch to honour the recent contributions made in Belgium at the Catholic University in Louvain by Van Hootegem and his staff [11, 12], in part via the Flanders Synergy Program [8].

Because the L-STSD-approach shares the systems theory as a foundation with Enterprise Engineering, and the other OD-approaches mentioned above lack this foundation, the jump to the conclusion that L-STSD may be easy to compare with EE and may also the most probable OD-candidate when it comes to enriching EE with a vision on organisational structure design seems allowable.

2.2 Model of an Organization

An organization implies work, to be done by members, often using machines and equipment, aimed at realizing a main goal or higher purpose, which is to supply something valued by customers, or to deliver a contribution to society. System theory models this as a system where input is transformed in the output to be supplied (Fig. 1).

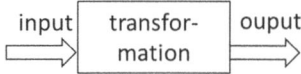

Fig. 1. An organisation as a system

Along with the main goal, an organization has several sub goals (stay viable, offering employment, pay tax, increase shareholder value), and it is using all kinds of resources (e.g. machinery, ICT, building, money, knowledge) when doing all the work involved in realizing the goals set. Most organizations are permanent, or want to stay viable and hence must adapt to the threats and challenges from their competitive and (rapidly) changing environment. Some organizations are temporary, and 'live' only during the time it takes to realize a unique result. We usually see this in the civil engineering and building industries. When the bridge or the skyscraper is finished, the temporary organization is dissolved. This means that the contractors that participated in the project take back employees and equipment, and assign these resources to a new project with its own temporary organization. Below the focus is on permanent, viable organizations.

2.3   Structure and Its Functions

A simple definition of organizational structure is 'the division of labour and the coordination (that results from this division)'. Based on that, all organizations, big or small, new or old, for profit or not, have a structure. Networks of organizational actors have a structure too, because the work is divided among the network partners (being organizations or self-employed individuals), and intra-organizational coordination mechanisms must be in place.

The labour to be divided in a given organization is varied in nature and involves all the activities that the organization has chosen to do in house (instead of outsourcing it), which are necessary to deliver value to the customers that place orders. It comprises the primary transformation or make activities, preparatory work linked to orders, and support and maintenance work for the primary activities. Together these three types of activity are grouped and labelled the Production Structure (PS). Regulating the primary transformation by the Production Structure is the function of the Control Structure (Fig. 2).

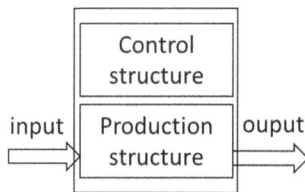

Fig. 2. Transformation involves a production and a control structure

The resulting need for coordination also equals work, usually, but not necessarily, done by managers. Some authors use 'coordination and control' as a fixed combination, which indicates that they see control as something related to, but not the same as coordination.

Ashby [2] distinguishes three layers in regulation: Control, Regulation by design, and Operational Regulation. Control, or 'strategic regulation', encompasses setting the goals of the organisation and monitoring whether the work that is done actually leads to reaching these goals and to take action if something happens that might prevent realization of the goals. Goals may be strategic, like entering a new market with a new product-service-offering, or operational e.g. meeting the revenue-target and reliability of order delivery. Then Regulation by Design develops changes needed to reach new strategic goals. Because these changes usually are referred to as innovation projects, this layer is sometimes referred to as 'Innovation Structure'. Innovative ideas may also come up when reflecting on operational activities or in contact with customers. This also leads to innovation projects to be carried out by the Innovation Structure. Finally the Operational Regulation of the Production Structure is the lower part of the Control Structure [1, 24] (Fig. 3).

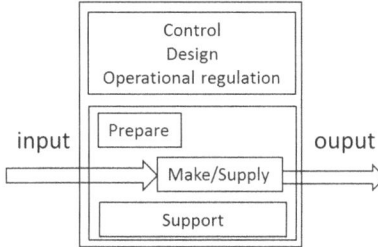

Fig. 3. Detailing the production structure and the control structure

So far, in spite of the term Structure, we dealt with functions, or contributions to the system, which must be fulfilled in every organisation that aims to deliver value, and stay viable. A self-employed professional will do all the PS and CS work for his one person organization. In the classic functional bureaucracies each kind of PS-activity is subdivided and assigned to an organizational unit, and a hierarchy of managers is needed for operational control and strategic management, with specialist staff in project teams for innovation work.

2.4 Contingency, Complexity and Challenges Shape Structure

An increasing size of the organization, as measured in number of employees and number of sites, will usually lead to an increasingly more elaborate, formalized, and in the end even really complex organizational structure. Contingent factors, like the dynamics of the market and technological development are to be considered when designing a structure. When demand is seasonal, there is a need to vary capacity that must be 'designed in' the structure to ensure a match between supply and demand. A fast growing demand and workforce implies frequent rethinking of the structure or a scalable type of structure where PS-units can be added easily.

New technology may enable automation of work and consequently leads to both a reduction of staff and to new kinds of work. So implementing new technology also demands rethinking the division of labour and always leads to changes in the structure. STSD in general stresses organizational choice, meaning that technology does not dictate the new structure.

New generations of higher educated workers are said to have higher demands on their job; it must enable them to use and develop their talents, and by a visible contribution to the organization's higher purpose also be meaningful for society. Employers should also respect their work-life balance, and work should not be a threat to their health and well-being. As customers they are also more critical.

Finally, the general public, ngo's and governments demand more sustainability and environmental awareness, social responsibility, and also, since the 2008 banking crisis, more ethical behaviour from organizations.

Taken together, the bar is raised for organizations, and it is hard for organizations with hierarchical, activity based, centralized, bureaucratic or matrix and multi-dimensional structures to meet the present standards of multiple value creation. This is caused by the fact that these kinds of structures require so much coordination between the highly interdependent units, that they are too complex to manage from a central point at the top of the multi-layered hierarchy. Coordination also involves work for highly payed managers, so it is costly too. And because all coordination and control entails information exchange, the information processing demand is much greater than the complex structure can provide, no matter how much ICT the organization employs.

Before explaining the basics of L-STSD, some characteristics of structure need to be highlighted first.

2.5   Division of Labour Influences Coordination Need

Organizations work, or rather, in organizations work is done by several people and part of the work is automated and done by (computer controlled) machines and ICT. Is it relevant and necessary to systematically design an organization or can structure be left to self-organizing? Adam Smith divided the work of needle makers into simple steps, and put workers in a row, Hundred years ago Frederic Taylor divided doing from thinking, which made us believe somehow that people that 'do' cannot 'think' and that thinkers better not 'do'. He also aimed to improve the way work was done by using scientific approaches to find the one best way of working. Implicit here is the assumption that when all individual jobs are scientifically optimised the whole will perform optimal. Closely related to Tayloristic thinking is the idea of an economic order quantity to spread the cost of changeover and inventory optimally over a batch of products, leading to a firm belief in economies of scale.

So we have over a century of history leading many to believe that the only logical ways to group work are the assembly line for mass production and else the activity based or functional approach, where ideal batch sizes are based on flawed economies of scale-thinking.

When organizational work is divided over several people, or organizational units, it must be coordinated to make sure that the result meets the request of the customer, and other norms and standards. The different contributions are usually interdependent, and it is easy to imagine that different divisions of labour can lead to different interdependencies, and hence to different needs and ways to coordinate. Some examples to show this. A small coffin of a child can be carried by one person, so no coordination, but an adult needs six carriers who must coordinate their movements when lifting the coffin and walking to the grave. When each craftsman makes, finishes and assembles parts for one piece of furniture, there are no interfaces between the craftsmen. When you group them in three departments, for making, finishing and assembling, these departments need to be coordinated.

Such sequential interdependencies between groups can be compared to a relay-team handing over the baton, which stands for the order to be processed. Most of the problems of any relay team have to do with handing over the baton, because at high speed it requires very precise coordination effort to get it in the hands of the next runner. Coordination means extra work, and it can go wrong, the falling baton, and according to Murphy: it will go wrong. Hence, designing a division of work with as little handovers or interfaces as possible, will reduce much of the coordination work, and so reduce both risk and the number of people doing that (staff, management, liaison roles, committees to name a few). At the same time less interfaces equals less interface problems, so less chances of disturbances, and the associate effort to solve these. Again, compare the 4×100 m relay team with a single athlete running 400 m. The latter may be a bit slower, but the chance of missing a handover leading to disqualification is zero.

Concluding, a functional division of labour that leads to interfaces where coordination is needed and where troubles find their cause, may not be such a good idea. A redesign forming independent units with far less interfaces may be smarter, and has another in-built advantage for the employees.

2.6 Division of Labour and Quality of Work

Without people organizations obviously do not exist, because even the 'unmanned factory', or the fully automated internet based system for processing of payments by banks, requires designers, builders to erect, and people watching its functioning, solving defects, and do maintenance. When you take the importance of employees, including managers, serious, their work should be of high quality. Is work in an activity based, high interface situation of high quality? What about sharpening the point of countless needles all day long (Adam Smith), shovelling tons of iron ore all day (Taylor's famous example) or making legs for chairs all day, and for tables the next day? Apart from the boring character of the repetitive work itself, in such structures someone else decides about the quality of your work, another tells you what to do, when to do it, and in how little time. Western educational systems do not deliver needle-point-sharpeners, ore shovelers or table leg makers to the labour market, but craftsmen and women with considerable more knowledge and skills than the illiterate farmer sons that Taylor and his contemporaries could hire for the work to be done at that time. Nowadays the organization designer should match the work assigned to organisational units to the higher skills of workers.

Following the previous example of furniture making, when in a group of ten workers, each is coupled with the complete set of tasks to make furniture, the possibilities to develop all necessary skills and become a real craftsman, are much better. Each craftsman then has the task to complete a table or cupboard on his own, maybe even including the contact with the customer about her demands for it. This grouping automatically facilitates job enlargement. So, instead of units that do one kind of activity for all orders, leading to monotonous, poor quality of work, without much opportunities for job-rotation within the group, (and the large number of interfaces), the designer should group work in such a way that units are able to do all work needed to fulfil their sub-set of customer orders; these units are a mini-factory or mini-service unit, that in principle can manage its own affairs.

Or, to put it differently, the structure shapes the conditions for meaningful self-management, and various team members can learn to perform internal and external coordination too, which is a form of job enrichment. This way the quality of work improves 'by design', and it goes along with the reduction of interfaces which improves controllability explained above. So this kind of production structure design for the primary transformation and its preparatory and support tasks, leads to a win-win design for employees (higher quality of work), customers (higher quality, more reliable performance with less chance of broken promises), and shareholders (coordination costs reduced, higher performance, higher margins and/or turnover).

It should be clear by now that a simple structure, built of units that are as independent as possible, has reduced coordination needs and costs, while at the same time offers higher quality of work for employees, because it enables learning on the job within the team and gives job control to facilitate meeting the job-demands. With such a structure teams can be(come) truly and meaningfully self-managing, like mini companies. Now the question is: how can you design or redesign such a simple structure?

3   L-STSD Design Principles, Sequence, Parameters and Function Models

3.1   Starting Points for (Re-)designing Structure for New Organizing

There are three different design situations. First is a start-up with growth potential that uses a philosophy that accommodates growth easily. Secondly when a big company starts a new plant or subsidiary where a considerable number of people will work shortly after the office or factory is build or rented, a new structure is designed and new people are hired that match the chosen design, maybe based on a new design philosophy. The third, and most frequent is in fact *re*design of the structure of an existing organisation. The latter is labelled a brownfield design, and the former two are called greenfield designs.

The essential difference is the number of constraints. In a greenfield situation it is easier to come close to realizing a 'castle in the air', because people, equipment, ICT and building can be matched to the ideal structure. The brownfield may suffer more from path dependencies, or all choices made in the past for equipment, building, or ICT that do not fit the ideal design, and act as a constraint. This means that equipment designed for economies of scale that fitted well in an activity based structure will have to be part of the new situation, and will create undesired interdependencies between units, at least for the remaining write-off period. Lean labels this kind of equipment 'monuments' or bottlenecks. All orders need to pass it, even though before and after it three independent flows could be formed. As soon as new smaller scale equipment is available, and/or can be afforded, the monument can be replaced by three, not necessarily identical pieces of equipment for each of the flows, to make these independent end-to-end.

The following explains the design chain and is based mostly on the latest L-STSD handbook by Kuipers et al. 'Het nieuwe organiseren. Alternatieven voor de bureau-cratie.' Translated: 'The new organizing. Alternatives for a bureaucracy' [15].

3.2 An Overview of Lowlands-SocioTechnical Systems Design

The handbook [15] prescribes a six-step design approach to organization design, and in the fourth step the organizational structure design is made. All designers know that such processes are in practice not sequential but iterative in character. However the order in which the design steps are presented follows from engineering or design thinking logic.

At a higher level, the six steps are part of the four-phase intervention in organization-cycle of:

diagnose – design – implement – evaluate.

Preceding the design phase, a diagnosis should be made in which it becomes clear that the current problems cannot be solved, nor the strategic goals reached, with the current structure. Alternatively, when there are a lot of problems with ICT, e.g. legacy, too many applications, complex interfacing, the solution may seem to implement a new ICT-architecture. However, then the structure should be scrutinized first, to see whether that is really (near) future proof, especially if there is reason to expect that changing the structure will be less easy after the new ICT-architecture, probably including an ERP-system, is implemented.

Following the design, it is implemented in the change phase. After the implementation is finished and the new structure and systems are settled somewhat, the project should be evaluated.

The six L-STSD design-steps are:

1. Choose the scope of the redesign; preferably the whole organization, because that enables the greatest reduction of interfaces.
2. Write a statement, if it is not available yet, in which the higher purpose and goals of the organization are explained, and mention the multiple values to be created, for the multiple stakeholders involved.
3. Detail the higher purpose, and the values to be created in functional design requirements, and make them as 'SMART' as possible, to serve as criteria to guide and judge the alternative design.
4. Design the organizational structure, starting with a top-down Production Structure design, followed by a bottom-up design of the Control Structure. This order is unique in the field of OD. The Production Structure will have groups as independent as possible, and the Control Structure will be as decentralized as possible.

5. Design the hard and soft systems to make them fit with the chosen division of labour in the previous step. This indicates that ideally, the systems (automation, ICT, building) should fit the division of labour that delivers the best quality of work, instead of adapting the workers to the technology used; it is indeed socio-technical design, not techno-social.
6. Preceding the implementation the behaviour, leadership style, culture, and 'the way we will work' must be redefined in a way that fits the purpose, functional requirements and structure of the new organization. From the difference between the old and the new situation an action learning plan to learn and train the new behaviour and style can be designed.

The logic of Steps 1, 2 and 3 will be obvious to the reader, but the others need some explanation.

3.3   Step 4a the Production Structure Design Is Top-Down

As mentioned above the structure (re)design starts with the production structure, and this is done top down. The aim is to increase controllability of the production structure by reducing the number of interfaces between groups as much as possible. To do that first the units and jobs involved in making must be found in the old structure. Then the products and/or services made, the markets served, and the various inputs are listed. A bit oversimplifying, you start with N employees 'making' and you must divide these first into independent units, lean would say 'flows', linked to a product group or to a part of the market (geographically or type of customer). If needed you 'cut' the branches further in smaller independent units, until you end up with groups that have a size that allows the members to feel and work like a team: 6 to 12 employees as a rule of thumb (with 4–20 as extremes). If possible you divide in flows all the way down, because those units have an end-to-end responsibility for their own subset of customer orders. Ideally these team can work parallel, and independent, without intensive coordination, only an occasional exchange, e.g. of knowledge or capacity when workload is low in one of the flows, and high in another with similar employees and work.

Then the attention is turned to the preparing activities and the employees that held these jobs. They may initially be placed as specialists inside the teams, because that reduces interfaces compared to making separate prepare groups that are linked to several make-teams. Only when there is a good reason, a separate prepare-group can be designed, and then an attempt is made to form mixed groups of staff linked to some of the (sub)flows.

Finally, the supporting activities are studied, with the same basic question as for preparing, whether some of the people can be transferred to the teams, or that, after training some of that work may be added to the jobs of the members of the make teams. Simple maintenance of team equipment, or cleaning the team's own workspace, makes them independent from a staff group that has its own ideas about the ideal moment to maintain or clean. For some very specialized or skilled tasks of a handful of specialists that must divide their attention to all teams, a central support group may be formed.

Placing all preparing and supporting activities (with staff) in the independent flow related make-teams is the theoretically ideal 'least interface' Production Structure design, where each team is a plant-within-the-plant, a hospital-within-the hospital, or an office-within-the-office. Of course there are also organisations that serve a large geographical area, where each sub region or neighbourhood has its own clients, e.g. home nursery care, or police, that may be served by a local team with a small office in that neighbourhood.

In practice this ideal of dividing in (sub)flows all the way down is not always possible. There can be reasons why only sequentially dependent units can be created at a particular design level or 'cut'. In an automobile plant there are usually three basic steps; making the body parts and weld or glue them together and link bonnet, hood and doors. Then the largely automated step of painting is done. Finally the painted bodies enter the assembly lines to become cars that are driven to the parking lot to wait for further transport. The painting equipment cannot be split in two, one for each assembly line. And the presses require a different kind of operators, and are universal and not related to the models that the assembly lines are dedicated to. In such a plant the three units are sequentially dependent, and L-STSD names those kinds of units 'segments', and their interfaces cannot be avoided. Within that constraint, the designer then tries to make units within each segment, that are flow-based, and the two assembly lines are an easy guess, but the body painting part will be less easy dividable in parallel flows.

As a simple example, when there are 24 employees making 'stuff', using the rule of thumb you may design 2 teams of 12, 3 teams of 8, or 4 teams of 6 employees (and maybe 8 teams of 4). Now, suppose you have to process two materials, metal and plastic, to make three products, for four sub-markets. This matches to the 2, 3 or 4-team options, material, product, or market based respectively. Then, figure out which option leads to the smallest number of interdependencies or interfaces between the teams. In such a small organisation, Production Structure design involves just one divide or 'cut'.

In bigger companies more divides are necessary, before the last cuts in each branch deliver the 6–12 (4–20) employee make-teams. A grouping of teams forming a business unit together should add up to a maximum of 150–200 employees according to Kuipers et al. (2010). The Dunbar number, related to the maximum amount of people humans can relate too, is around 150 people. W.L. Gore is dividing its plants when they grow beyond that number, and builds a new one 'at the other side of the parking space' when the area allows that, of course. When the designer has to deal with an about 500 employee organisation (all working in one site), a first cut tries to find three independent units of about 170 employees.

There is no general rule of thumb on how to divide the Production Structure of a big multinational, selling different product & service categories, produced in numerous plants throughout the world, but basically the question always is: how can we make the units as independent as possible.

A separate kind of design challenge occurs when demand is local, and must be supplied by local small units; like fast-food e.g. McDonalds, Starbucks, KFC, with their large number of outlets, including franchises, or home care e.g. Buurtzorg Nederland with their independent, self-managing neighbourhood teams serving clients in 'their' neighbourhoods, with other teams in surrounding areas.

This top-down design procedure and rule is very similar to modular product or system design and in fact uses the hierarchical decomposition described by Herbert Simon. And evidently, a modular product design enables the forming of Production Structure groups that are responsible for their own module, and an assembly group to combine the modules each customer has chosen. This may be a PS with two segments (modules and assembly), and sub-flows in the module-segment.

So far we have been dealing only with activities that are necessary to fulfil customer orders by preparing and making, and supporting. The resulting set of PS-units needs to be coordinated, or 'pasted', when they are interdependent, and control is always needed to make sure everything works according to the standards set, so that customer expectations are met (or exceeded), in spite of the disturbances that occur. For this the Control Structure is needed, and only now it can be designed, bottom-up.

3.4   Step 4b the Control Structure Design Follows Bottom-up

The control structure entails three different types of control, and they have a logical hierarchical order. They are named, bottom to top: operational regulation, regulation by design and strategic regulation.

When the design of the production structure is ready there are a number of units that are as independent as possible. At the lowest PS- and CS-level each unit needs operational control of its own internal processes. Then, for the units that have an interface with other units, an inter-unit second layer of operational coordination and control must be designed. Depending on the size of the organization and the design of the production structure, an operational control hierarchy is designed. Do note that these layers are not the same as hierarchical layers depicted in the orgchart. The top layer just ensures that all coordination and potential synergies between PS-units are taken care of.

The operational control activities are triggered partly by disturbances in the external environment, partly by problems with an internal cause, often at the interfaces. Which employees are charged with operational control activities is the second question. They may be assigned in a classical way to first line supervisors, and higher managers, but the autonomy of the units creates better opportunities to assign much of the operational control activities to one or more members of the unit.

Then next layer, named 'regulation by design' based on Ashby [2], can be designed. It is also named 'innovation structure'. This entails all the innovation and change activities that an organization undertakes, which are usually done as 'projects' and 'programs' or 'portfolio's' (sets of related projects). Before a project can start, opportunities have been searched, business cases or project proposals drafted for the most promising opportunities, and a selection of the best project proposals is then made by an 'innovation portfolio board', based on strategic criteria. Considering who should search, make business cases, select them, and be part of the project teams, is again a design question. In a functional structure separate groups of innovators exist. And the stage-gate-approach leads to intermediate project results being 'thrown over the wall' to the next functional group at each gate, until the result is finally received by Operations to make, and by Marketing to sell. And it is regularly thrown back too, when the downstream group finds flaws that must be corrected by a previous unit.

Concurrent engineering already tried to solve this by bringing the downstream functional representatives in the project from the earliest stage of the project on, to make sure that all relevant perspectives could be taken into account. In these cases innovation project teams are a mix between fulltime innovators, and part-time contributors that spend most of their working hours in PS-work and/or operational regulation. The former may be an industrial design engineer, a service developer, and the latter may be from manufacturing, logistics, quality control, to help meet standards for manufacturability, simplify logistics, improve quality level adding statistical process control thinking to the design team. Laloux [16] presents an example of FAVI, a French company that goes further in decentralizing innovation tasks. There a machine operator may visit a trade fair, see a useful new machine, and be appointed to the process innovation project managers role by his team to carry the innovation case further. When the new machine is up and running he comes back in his operator role.

The top layer of the Control Structure is named strategic regulation and that involves defining purpose of the organization, the values to be created, and the goals to be reached in the short and longer term. Following from the central definitions of the goals, these must be deployed to (when applicable) the divisions, their strategic business units, and their business units, until the floor units are reached. In this process each level clearly sees its contribution to the purpose of the organization as a whole.

The basic design ideas of independent units and decentralising control responsibilities, implies that decisions on local strategy and innovation that are only influencing the particular unit on the shop or office floor, should be made by members of that unit, when needed supported by specialist staff. When an innovative idea relates to a business unit (here a group of related floor units), then that project 'belongs' to the business unit and floor unit members and specialist innovator staff may work together in the project team. Radical or disruptive innovations usually lead to a new (business) unit in the production structure, and must be coupled to the corporate level because no (business) unit can 'own' it, and even may regard it as a threat to its current market position.

3.5   The U-Shaped Design Order of L-STSD Is Unique

Lowlands STSD is believed to be unique in its systems theory based design sequence, and its aim to improve controllability of the organizational system by designing independent and self-managing units. Because these units are made responsible for an output that is recognizable as a meaningful subset of the total output of products and services, all work in the production structure is firmly linked to the purpose of the organization. Because a unit fulfilling a complete customer order on its own will do a variety of tasks, the potential for job enlargement is close at hand for each member of the team.

The independency of the units is a precondition to assigning various control tasks to at least of part of the members of each unit, enriching their jobs. There is a much greater variety of coordination and control tasks that such a unit can do without interfering with other units, compared to units in traditional, activity based structures. The latter are so tightly coupled to many other units in pooled, sequential, and sometimes even reciprocal interdependencies, that almost all control decisions need to be coordinated with other units, and only a small subset can be self-controlled. Giving such a unit in a functional structure the name 'self-managing team' is misleading the members, and in fact wrong from the systems theoretical perspective explained above. Compare it with a 'team' of 3-year olds that believe that they are self-steering the police car to the crime in the Merry-go-round, the structure. Each seat has its own wheel. They soon enough find out that no matter how much, or in what direction they turn their wheels, the car just goes round in the circle dictated by the design of the Merry-go-round.

Based on the systemic or cybernetic logic the Lowlands-STSD design sequence is top-down for the production structure, and then building the control structure bottom-up. When controllability is the ultimate design aim, this sequence is believed to be the only way to optimize the design of the organizational structure; reducing complexity implies less disturbances, decentralizing control implies faster response, and in combination the quality of work is increased because both job demands and job control can be balanced by the units themselves. In other words L-STSD uses economies of flow or scope instead of economies of scale.

Now that the PS and CS are designed and people are placed in teams, it is about time to remember that organizations are not just people, but 'sociotechnical' systems. The people use all kinds of technology and knowledge, which are either hard (building, equipment, ICT, vehicles) or soft (quality system, handbooks, procedures). So now, the technical hardware deserves attention.

3.6   Step 5 Designing the Hard and Soft Systems Needed

When the design of the structure is completed in Step 4, the systems must be (re)designed in such a way that the units have all means they need to fulfil the assigned work. L-STSD proposes the 'First organize, then automate' as the leading principle, so that's why Step 5 deals with it. It is worth noting that the systems can be distinguished using the types of work described above:

- Production Structure systems:
  - transforming systems, like CNC-machines in CAM-systems or robots, or IT-systems for the workflows in data-processing companies
  - preparing systems; for all kinds of preparatory activities
  - supporting systems for the support tasks
- Control systems for
  - operational control
  - innovation
  - strategizing

Some of these systems are stand-alone and dedicated to one activity, while others are from different vendors, but linked or integrated (CAD to CAE and CAM/CAx into CIM), and some (like ERP) are combining various functions in one (modular) package.

When redesigning an organization there is usually a lot of existing hardware and software. In manufacturing big machines based on economies of scale-logic will not fit the smaller units. Buying new and smaller equipment must be both possible, and affordable. ERP-systems are adaptable to any structure via templates and parameters, but the step from an activity based structure to a flow based one usually implies a new implementation with all associated costs. This may even stop the restructuring. Alternatively, the new structure may be so simple that there is no need any longer for such a complex ICT-system, and ERP becomes obsolete [10, 19].

3.7 Step 6 Soft Factors like Behaviour, Leadership, and Culture

For the sake of completeness, the sixth step must be briefly mentioned. When employees from different departments are brought together in a group, it will not instantaneously be a team. Maybe used to just do what you're told, shifting towards taking more initiatives, and learning the other tasks in the team also takes time.

And managers and supervisors who were used to a command and control style and centralized authorities, must adapt to the situation where as much responsibilities for control as possible are decentralized to the group members of truly self-managing teams.

Depending on the starting point, it may take at least about two year or more, according to Kuipers et al. (2010) before the organization will work in a completely self-managing mode. Patience with this process with higher management and the Board may be one of the most important conditions to make such a holistic change a success. As far as the approach to change is concerned, it is common knowledge that going from a state A to a new state B, in which other behaviour will be the norm, can only be accomplished using a B-method. Using an A-method here would be like making people self-managing on command.

3.8 Diagnosing and Designing a Structure Using Parameters

Before and during a design project the current structure should be diagnosed, and the new designs should be judged respectively. To do that L-STSD provides the parameters developed by de Sitter [1, 15, 24–26].

De Sitter's parameters aim to capture where the structure or the design is between a maximal degree of functional organizing (i.e. activity based units with lots of interdependence and operations separated from control) or a minimal degree (i.e. independent, autonomous units delivering value to a subset of customers, with as little central or shared units). There are seven or eight parameters, depending on the source, and it seems beyond the scope of this paper to present all in detail.

Low parameter value structures are better in terms of their controllability, and of the higher quality of labour that come with the production structure with independent units and a decentralised control structure.

3.9 Diagnosing and Designing a Structure Using Function Models

When redesigning the focus may be automatically on the work already being done 'somewhere' in the old structure. But some systemic function may be missing, and that can cause problems too. To find these missing functions the Viable System Model (VSM) developed by Beer [3, 4] is a useful tool. It is comprising five, interlinked functions that together are 'necessary and sufficient' for viability. According to Beer, not fulfilling a function of his 'viable system model' (VSM) adequately, or not link them together, will eventually lead to problems with the viability of the system. So far the rather bold claim by Beer about the functions being 'necessary and sufficient' is not convincingly challenged. As an example, if nobody searches the relevant environment for signals that imply that something in the offering of the organization must change, e.g. a truly disrupting innovation being developed by a new competitor, it will only notice it when sales are rapidly declining.

Because of the very abstract nature of the VSM, the 'Model Innovation and Organisation Structure' (the MIOS) was developed by Lekkerkerk [17, 18] as a descriptive and diagnostic tool for structures, with an emphasis on the 'innovation structure' (or 'regulation by design'). The VSM and the MIOS enable answering the question: 'Are the functions carried out and are they properly linked, or not?' If not, they must be added or linked. And, when carried out, are they formally assigned or just informally done? When employees perform a function because they think they should, or because they 'just like it', this may sound fine and proactive. But when it is not in their formal job or role descriptions, this function stops when they move to another job, and that will inevitably lead to problems.

3.10 Summarizing L-STSD

Combining the specific design order, PS top to bottom and then building CS bottom up, and 'systems follow structure', should lead to better structures. They have low parameter values and all VSM or MIOS functions are assigned.

It is interesting to note that some organizations developed structures that can be called 'Lowlands sociotechnical' without any apparent knowledge of this specific Dutch approach. So the kind of systemic logic to build a structure with can be found or reinvented without knowing L-STSD. But using this theory-based approach will lead to a 'good' structure faster. A few examples follow.

Ricardo Semler, the Brazilian entrepreneur, applies it in his company Semco, and his book 'Maverick' is widely read, and he is often invited as an inspirational speaker at expensive business conferences.

Ten of the twelve inspiring cases in the book 'Reinventing organizations' by Frederic Laloux [16], show these kind of structures too. Laloux also presents two Dutch cases, BSO and Buurtzorg Nederland. BSO was founded in 1973 by the late Eckart Wintzen, so before De Sitter started publishing with his group. He used a cell division approach. As soon as BSO-units grew beyond 50 employees, they were forced to make a plan to divide in two, which would be executed as soon as the headcount was around 60–65 people. To the cells he added only very limited central units for support at headquarters. Buurtzorg Nederland (BN), the other Dutch example Laloux presents, was started near the end of 2006. Its founder, Jos de Blok probably learned about De Sitter's work in a reorganization at his former employer. By 2016 BN employs some 900 teams of about 12 nurses, who deliver care to the clients in their neighbourhood. Headcount in The Netherlands after only ten years is around 12.000 employees, of which only about 40 work at the central support office.

Now the basics of L-STSD are explained, by summarizing mainly Kuipers et al. [15], the reader who is familiar with Enterprise Engineering may understand these basics to such an extent that the following comparison between the two now can be appreciated.

4 Comparing EE and L-STSD

As said, the comparison is based on a limited set of EE-texts [6, 7, 13, 14, 20], and a comparative description of Enterprise Architecture methods [23], and some were a bit hard to understand at first, so this is far from the final word, but rather a first attempt to compare and contrast.

4.1 Observed Commonalities

Apart from the similarities listed in Table 1, there are at least four more. In the attempt to use a holistic approach EE resembles L-STSD, because that approach rather goes for an integral redesign of a whole organization. Redesigning a sub-system may sometimes be inevitable due to pragmatic or power reasons, but it will probably lead to sub-optimization because of the relations between the rest of the organization and the unit to be redesigned.

Secondly, both EE and L-STSD find common ground in their use of general systems theory as a basis.

The fact that EE regards people as the 'pearls' of an organization, and wants to improve organisational functioning ('reducing strategic failures'), resonates well with the L-STSD aim to enhance Quality of Work (for employees), while at the same time improving Quality of the Organization, under which heading mostly factors related to external stakeholder value creation are comprised.

Fourth and finally, Herbert Simon's modular or hierarchically decomposable systems are mentioned in the EE-Manifesto [6] and are also clearly recognizable in the top-down Production Structure design approach. That a modular design of an organisation or a system would be based on atomic elements, as The Manifesto states, seems a first difference.

4.2 Differences

The Manifesto states that modular designs are based on atomic elements. Elsewhere transactions are defined as these basic building blocks [13, 14]. When atomic elements, the transactions, are put first, linked together in chains to form a process, the question comes up how to link all processes to the organizational structure. This has to do with the L-STSD-idea of interfaces and aiming at their reduction. A process that involves multiple departments, must have different transactions for each department. Let's take a furniture example with four functional groups: (1) a request to make parts, (2) one to paint parts, (3) one to assemble them, and (4) another to transport and install it at the customer's house. However, when a firm making the same kind of furniture employs craftsmen who deal with all the four steps, then one request (or transaction) seems sufficient: 'make a drawer for Mrs. Hofer at Farmer street 24'.

A related question is when to stop dividing transactions further into sub-transactions. This surely is a problem for authors of procedures and (quality) manuals for an organization. If everybody in the organization knows where to get coffee, do you need to divide the request 'Please get me a cup of coffee.' into al the steps needed? However, if the organization has a complicated manual espresso machine with lots of controls and buttons, the situation is different.

It should be investigated how the L-STSD-rule of 'minimum critical specification' is different from the apparently highly detailed EE-idea of atomic transactions, and perfecting each. And what to do with transactions that are not linked to an IT-system is puzzling me.

Another problem linked to (my understanding of) the atomic transactions, and the idea of perfecting all to build a perfect organization, is that it looks like something that Taylor advocated, wrongly assuming that optimizing each sub-task lead to optimal processes adding up to an optimal organisation. Dietz et al. mention that Taylor was heavily criticised (p. 90 [7]), and seem to agree, so why use his wrong idea.

Also, Dietz et al. [7] remark: "Obviously, an effective design science must have its fundamentals in the natural sciences." When it comes to technical systems, governed by laws of physics, as are designed by architects, civil, mechanical, aeronautical, and electrical engineers, to name a few designers, this seems right. However, the statement denies the fact that 'knowing how' may do for designers making design choices, although they may regret that they are not yet 'knowing why' precisely. These 'rules of thumb' that are developed from 'what worked well' can be proven, and usually much refined, by empirical research in one of the natural sciences. An example may be metal fatigue and the 'Wohler-curves'.

Furthermore, it is highly questionable whether a complex system like an organisation, consisting of human 'pearls' (and much else), will obey laws from the natural sciences in a deterministic way. Its interactions with its complex environment have so many unknown factors, and consequently the outcomes of actions, based upon decisions may never be known for sure in advance. The book by Achterbergh and Vriens [1] has a title explaining this: "Organizations. Social systems conducting *experiments*." The stress Dietz et al. [7] put on theoretical underpinning of enterprise design seems to negate the idea of bounded rationality, and the tendency of people (which managers obviously are, in spite of the super-human nature ascribed to people like Jack Welsh, Steve Jobs, or Michael Dell) to 'satisfice' (Simon), because of time and knowledge constraints, rather than to 'optimize', in the many and complex problem solving and decision making situations they are facing every day.

Also, managers and social scientists will simply not believe that there is a one best way to engineer the entire enterprise in every detail, and if it would be possible, then the design is already obsolete the moment it is finished, because detailed design is so time-consuming that the world around already necessitates another design before it is half way done. Because of this, sociotechnical designers advocate and use the idea of 'minimum critical specification' when designing. Basically this means that you design only the key characteristics and leave ample room for employee self-design and self-control to take care for all eventualities and disturbances that cannot be incorporated in a formal design. Designing all atomic transactions, seen as the building block of the enterprise, seems to violate this 'law' of minimum critical specification.

As far as the three generic goals of EE are concerned, there is little doubt that EE and L-STSD agree upon social devotion. For intellectual manageability, it appears that EE eventually hopes to develop theory that may master the complexities of the enterprise. L-STSD rather argues that complexity in the environment is given, and that the only way to deal with that is to reduce the internal complexity, meaning that each dedicated organizational unit deals with a limited part of the environment only, and is internally as simple as possible by stressing the design of flows (with sub-flows, sub-sub-flows) until the unit has a headcount that allows it to function as a team. Chaos theory seems to tell us that an organization and its behaviour can never be fully determined because of their non-linearity and sensitivity to initial conditions. The third, organizational concinnity, seems to deal with the control structure, which L-STSD sees as the tool to link the production structure units into a coherent whole, but is not that clear.

Dietz et al. state: "Because of its holistic, systemic, approach, it resembles systems engineering [...]. But it differs from it in an important aspect: enterprise engineering aims to do for enterprises (which are basically conceived as social systems) what systems engineering aims to do for technical systems." [7, p. 92]. However, when designing both technical and social systems the people involved must always be taken into account, whether they work with the technical system, or work in the social system. So, where is the difference?

Finally, another potential difference is the order in which changes must be designed and implemented. L-STSD is explicit in prescribing to first design the division of labour, and only after that design the systems that match the structure, including the ICT that EE seems to care mostly about. This order should prevent the quality of work to be hampered by systems that enforce an implicit structure upon the organisation. Especially ERP-systems have a bad reputation among L-STSD practitioners for their lack of adaptability to changes in the division of labour that are deemed necessary after ERP has been implemented [10, 19]. If there is no such sequence in EE, it may be possible to start redesigning transactions and (re)build or refine ICT-systems, without dealing first with the question whether the existing division of labour, the Production Structure, is as simple as possible and sometimes the question whether it fits the new strategy that is currently under development is also relevant. Developing ICT for an obsolete structure, and may be even blocking the introduction of a new one due to the costs of ICT-changes associated with that, cannot be acceptable when EE aims to help (people in) enterprises work as effective and efficient as possible.

5   Conclusion

Although there are a number of differences between Enterprise Engineering and the Lowlands SocioTechnical Systems Design-approach, based upon this first comparison it seems that these can be solved.

A first point is the design sequence 'systems follow structure', which L-STSD favours. This seems to fit with EE.

A second point, the modular top down design of the Production Structure, which basically defines who needs what information to make sure that the coordination of interfaces works well, will probably not be in contradiction with EE.

Thirdly, the focus on atomic transactions needs to be rethought. When the basic (Production) Structure seems to be 'future proof', it can be useful to make all the details explicit and clear between al transaction partners. However, when a big change and investment in ICT systems seems necessary, e.g. to replace legacy systems, it is important to answer the question about the viability of the existing (Production) Structure first, and make the 'systems follow structure'.

It may be added that both EE and the founders of L-STSD around De Sitter like to use their own jargon or give their own meaning to terms that others use in a quite different sense. This is confusing and leads to managers putting aside their ideas as being too difficult, academic or theoretical to be of any practical value.

Both communities are also relatively small, not widely known, and somewhat closed like an oyster. However, they can both be regarded as pearls in the Enterprise Architecture and Organisational Design fields respectively. So, to help organisations function more effectively and efficiently, let's join forces, grow the pearls and open the oysters.

Acknowledgements. The authors thanks H.A. Proper for his invitation to be part of the special session on Organisational Design at EEWC2017.

References

1. Achterbergh, J., Vriens, D.: Organizations; Social Systems Conducting Experiments. Springer, Heidelberg (2009)
2. Ashby, W.R.: An Introduction to Cybernetics. Wiley, New York (1956)
3. Beer, S.: The Heart of Enterprise, 'The Stafford Beer Classic Library'. Wiley, Chichester (1994). (1st edn. 1979)
4. Beer, S.: Diagnosing the System for Organizations, 'The Stafford Beer Classic Library'. Wiley, Chichester (2000). (1st edn. 1985)
5. Burton, R., Øbel, O., Håkonson, D.D.: Organizational Design: A Step-by-Step Approach, 3rd edn. Cambridge University Press, Cambridge (2015)
6. Dietz, J.L.G.: Enterprise Engineering. The Manifesto (2011) (3 April 2017). http://www.ciaonetwork.org/publications/EEManifesto.pdf
7. Dietz, J.L.G., Hoogervorst, J.A.P., et al.: The discipline of enterprise engineering. Int. J. Organ. Des. Eng. 3(1), 86–114 (2013)
8. Flanders Synergy: http://www.flanderssynergy.be/
9. Galbraith, J.R.: Designing Complex Organizations. Addison Wesley Publishing Company, New York (1973)
10. Govers, M.: Met ERP-systemen op weg naar moderne bureaucratieën? eigen uitgave, proefschrift KUN (2003)
11. van Hootegem, G., Benny, C.: Slimmer zorgen voor morgen. Het nieuwe organiseren in theorie en praktijk. Acco, Leuven (2013)

12. Huys, R., Maes, G.: Geert Van Hootegem: Meten en veranderen. Instrumenten bij het nieuwe organiseren. Acco, Leuven (2014)
13. Janssen, T.: Werken aan samenwerking. Naar effectieve organisaties met Enterprise Engineering. Scriptum, Schiedam (2015)
14. de Jong, J.: A method for an enterprise ontology based design of enterprise information systems. Ph.D.-thesis Delft University of Technology. SIKS-dissertatiereeks 2013-39 MPrise (2013)
15. Kuipers, H.: Pierre van Amelsvoort, Erik-Hans Kramer: Het nieuwe organiseren. Alternatieven voor de bureaucratie. Acco, Leuven (2010)
16. Laloux, F.: Reinventing Organisations. A Guide to Creating Organisations Inspired by the Next Stage of Human Consciousness. Nelson Parker, Belgium (2014)
17. Lekkerkerk, L.J.: Innovatie- en OrganisatieStructuur. Ontwikkeling en test van een functiemodel voor structuuronderzoek en -diagnose. Proefschrift Radboud Universiteit Nijmegen, Innovatica, Nijmegen (2012). http://repository.ubn.ru.nl/handle/2066/93601
18. Lekkerkerk, L.J.: Verbinden van Organisatie- en Architectuur-ontwerp: een innovatie? Paper presented at EAM 2013 – Innovatie door verbinding (2013)
19. van Lieshout, T.: Simple and effective, designing information systems for modern organizations, Ph.D.-Thesis KUN-NSM, Nijmegen (2002)
20. Magalhães, R., Proper, H.A.: Model-enabled design & engineering of organisations and their enterprises. Int. J. Organ. Des. Eng. 1(1), 1–12 (2017)
21. Mintzberg, H.: Structure in Fives; Designing Effective Organizations. Prentice Hall, Upper Saddle River (1993). (1st edn. 1983)
22. Mohr, B., van Amelsvoort, P. (eds.): Co-creating humane and innovative organizations. Evolutions in the practice of socio-technical system design (2016). https://www.createspace.com/5593720
23. Santema, A., van Gils, B., Oord, E., Driel, M., van Rijn, R.: Wegwijzer voor methoden bij enterprise-architectuur, 2e druk, Ngi-Van Haren Publishers (2013)
24. de Sitter, L.U.: Synergetisch produceren. Human Resource Mobilisatie in de productie: een inleiding in de structuurbouw, Van Gorcum, Assen. Out of print (2000). (1st 1994, 2nd 1998/2000)
25. de Sitter, L.U., Den Hertog, F., Dankbaar, B.: From complex organizations with simple jobs to simple organizations with complex jobs. Hum. Relat. 50(5) (1997). doi:10.1023/A:1016987702271
26. Ulbo de Sitterkennisinstituut: http://www.ulbodesitterkennisinstituut.nl/

A Literature Review of Coordination Mechanisms: Contrasting Organization Science and Information Systems Perspectives

Maximilian Brosius[✉], M. Kazem Haki, Stephan Aier, and Robert Winter

Institute of Information Management, University of St. Gallen, St. Gallen, Switzerland
{maximilian.brosius,kazem.haki,stephan.aier,
robert.winter}@unisg.ch

Abstract. Information systems (IS) research has long been promoting the necessity of aligning local IS investments in organizations with their enterprise-wide objectives. One of the prominent means to realize such an alignment are mechanisms that coordinate various stakeholders in different organizational entities. Despite its prominent origins and manifold translations from organization science (OS), there is no single theory on coordination. The research at hand conducts a literature review of the underlying coordination mechanisms to offer a comprehensive understanding of coordination for prospective IS research. To this end and structured in eight categories of mechanisms, we contrast the reflection of coordination in OS and IS research. In outlining implications for future research, we also discuss how IS studies follow and complement OS research.

Keywords: Coordination · Coordination mechanism · Literature review · Organization science · Information systems research

1 Introduction

An increasing number of information systems (IS) change and development endeavors focus on creating local solutions for specific business needs [66]. Prominent reasons refer to organizational landscape complexities, time dependencies, and economic efficiency arguments that have led organizations to allocate IS change and development responsibilities as well as project ownerships to local business units [56]. While this allocation has brought about high performance gains on a local/short-term basis, in the long-run organizations have begun to face challenges in consistently aligning, integrating, and managing their corporate IS landscapes [52].

Over the past decades, IS research and practice have broadly addressed the necessity to *coordinate* IS change and development endeavors on an enterprise-wide basis in order to meet enterprise-wide and long-term intentions. In this vein, a particular group of enterprise-wide IS management approaches has been promoted, such as enterprise architecture management [67], project portfolio management [14], or enterprise application integration management [42]. Nonetheless, in response to increasing complexities and uncertainties, a key characteristic of many enterprise-wide IS management

© Springer International Publishing AG 2017
D. Aveiro et al. (Eds.): EEWC 2017, LNBIP 284, pp. 220–233, 2017.
DOI: 10.1007/978-3-319-57955-9_16

approaches is their operationalization under top-down, strictly governance-based coordination mechanisms (e.g., hierarchical authority command, control) [23, 24]. Notwithstanding their specific utility, these top-down coordination mechanisms reflect only a limited facet of coordination, and may thus not be effective in every organizational context [6, 24]. Hence, a comprehensive understanding, reflecting the magnitude and diverse facets of coordination and its constituent mechanisms, becomes necessary.

Coordination is a well-established research topic that has arguably been developed in organization science (OS) and later on adopted by IS research [7, 16]. It is defined as the achievement of "concerted action" whenever actors (e.g., employees) become dependent on one another, for example due to sharing the same tasks, resources, or goals [75]. In this vein, coordination is further defined as the "linkage and integration of different parts of an organization" toward a certain goal [77, p. 322]. Coordination becomes realized through a diverse set of mechanisms [45], which are defined as "tools for achieving integration among different units within an organization" [49, p. 490]. Mechanisms address specific and general problems of emerged dependencies [12] and "permit coordinated action across a large number of interdependent roles" [25, p. 28].

Despite coordination's manifold adaptions, translations, and interpretations from the OS literature, there is not a single theory on coordination [29]. Recognizing this general laggard, in the paper at hand we aim at contributing toward a comprehensive understanding of coordination for prospective IS research. As coordination has originated and been largely adopted from the OS literature, we provide an overview and contrast the existing research on coordination mechanisms in both the extant OS and IS literature. Furthermore, we emphasize a structured representation of mechanisms as a basis for guiding prospective IS research through the lens of coordination.

The paper at hand is structured as follows: In the next section, we present our literature selection and analysis method. Then, the results are presented in the subsequent section. Finally, we critically discuss the resulted insights along with a conclusion on further steps, implications, and limitations.

2   Research Method

In order to provide an overview on the phenomenon of interest, we opted for a review of prior research to identify the main discourses on coordination mechanisms in OS on the one hand, and to compare them with the discourses in the IS literature on the other hand. Both disciplines, OS and IS, have been discussing coordination for decades. A significant number of publications arrived in the top journals of both disciplines. Because of the large number of available publications and as we expect the highest quality work in the top journals, we limited our review to these journals. Hence, we selected the relevant peer-reviewed publications from both the *AIS senior scholars' basket of journals*[1] and a selected basket of highest ranked journals provided in *Harzing's journal*

---

[1] European Journal of Information Systems, Information Systems Journal, Information Systems Research, Journal of the Association for Information Systems, Journal of Information Technology, Journal of Management Information Systems, Journal of Strategic Information Systems, MIS Quarterly.

quality list[2] [30]. As a search strategy [78], we used the inclusion criterion of "coordi-nation" solely, searching on the EBSCOhost databases for (i) title (TI "coordination") and (ii) abstract (AB "coordination" NOT TI "coordination") fields. This was followed by a significant exclusion of articles (focusing hereby in particular abstracts, keywords as well as the main text body of the respective publications). Articles were excluded that either did not exclusively focus the linkage and integration of corporate units or stake-holders toward a certain ends [see also 44, 76] or did not have an explicit focus on coordination mechanisms [see also 48]. We used only the search term "coordination" because our study aims at providing an overall analysis of explicit coordination litera-ture, thereby also identifying different topics and discourses related to coordination mechanisms. In order to ensure the inclusion of influential and frequently cited publi-cations outside the senior scholar basket, we used forward and backward searches [78].

Following the suggestions of Webster and Watson [78], we developed a framework for guiding the literature analysis and for classifying the collected publications based on their topical focus of discussion. We built our analysis framework on the taxonomy of coordination mechanisms suggested by Martinez and Jarillo [49], who differentiate formal (departmentalization, de-/centralized decision-making, formalization/standard-ization, planning, control) and informal (lateral relations, communication, socialization) classes of mechanisms (Table 1).

Table 1. Overview of coordination mechanism coverage in IS and OS literature

Discipline	Hits	Hits after exclusion	Departmentalization	Decision-Making	Formalization/ Standardization	Planning	Control	Lateral Relations	Communication	Socialization
			Mechanism Coverage							
			Formal					**Informal**		
IS	146	30	10	5	4	5	5	8	4	6
OS	835	31	7	7	8	3	8	4	7	5
Total	981	61	**17**	**12**	**12**	**8**	**13**	**12**	**11**	**11**

As opposed to other forms of classification [16, 25, 29], this taxonomy offers an explicatory basis for a comprehensive review of coordination mechanisms. Since organ-izations maintain complex structures, different levels of functionalities, as well as vertical and horizontal integration, the developed analysis framework based on Martinez

---

[2] Academy of Management Journal, Academy of Management Review, Administrative Science Quarterly, Journal of Business, Strategic Management Journal, Management Science, Organ-ization Science, Organization Studies, Organizational Behavior and Human Decision Processes.

and Jarillo's [49] taxonomy gives an exhaustive abstraction of coordination mechanisms. Furthermore, it explicitly differentiates between formal and informal (more personal) coordination modes, which has been emphasized by early coordination mechanism literature [77]. Building upon this framework, we developed a coding scheme to systematically synthesize the collected publications.

As Table 1 indicates, prior to exclusion, coordination mechanisms have received different levels of attention from OS and IS scholars. This is partially due to the fact that the IS literature started comparably late addressing coordination mechanisms. Other reasons may be found in different foci of OS and IS discussing the respective coordination mechanisms. Due to our significant exclusion, we were able to not only filter those articles with the most exclusive focus on coordination, but also to select those with the most explicit coverage of coordination mechanisms. This led to a reduced and more compact set of articles for further analysis, given the initially large number of publications in the extant OS and IS literature.

The next section provides an overview on each coordination mechanism as well as on the main topics discussed in OS and IS for each of these mechanisms. We abstract the discussed topics in each discipline first, before reporting their main findings in order to contrast differences and similarities in OS and IS, respectively. Since the collected literature includes publications that address more than one coordination mechanism, some publications were assigned to more than one category.

3   Results

3.1   Formal Mechanisms

Departmentalization.   Often reflected in the organizational structure [49], departmentalization is discussed as a coordination mechanism to enhance business unit integration and horizontal work process alignment [26]. The corresponding linkage between departmentalization and coordination effects receives strong support throughout OS literature [26, 41]. Departmentalization has been mainly discussed in the context of organizational process design [47] as well as organizational unit segmentation [41, 79]. Early contributions reveal organizational process design as a coordination mechanism that impacts departmental interactions [7, 54]. With the same token, organizational unit segmentation has been introduced as a coordination mechanism that manages dependencies among work processes [46, 47, 63], and is impacted by the degree of integration and alignment as well as by the form of communication [29, 79].

IS literature mainly promotes departmentalization in the form of governance bodies, structural overlays (e.g., roles, groups) and physical colocation [7]. It refers to the extent to which business and IT entities engage in workflows, tasks, and processes related to IS and information technology (IT) functions [2, 28]. Furthermore, IS literature emphasizes the role of IT as a means to support departmentalization-related coordination [5, 22, 24, 70]. This role has been demonstrated in decreasing transaction costs of coordination [59, 60, 82] as well as in increasing task-related interactions [77] and communication [50].

Decision-Making. According to Martinez and Jarillo [49], decision-making can be reflected in either centralized (higher levels of command) or decentralized (lower levels of command) coordination mechanisms. The centralized decision-making is explained mainly by the structural design of the organization, particularly by hierarchies in OS literature [21, 55]. In the context of decentralized decision-making, social interaction and organizational communication [10] become increasingly important as mechanisms of coordination [10, 23]. The more decentralized organizations are, the more complex decision-making becomes [29]. This complexity is also expressed by the comparably high efforts of communication made in decentralized structures [33, 40].

The IS discipline investigates decision-making in decentralized forms, aimed at integrating business unit collaboration, as well as in centralized forms, aimed at mitigating risks of uncertainty [7, 81]. Thereby, decentralized decision-making promotes the role of individual responsiveness, for instance on IT systems development [16]. Drawing from the results of centralized decision-making, IS literature acknowledges the role of IT as supporting mechanism on enterprise-wide coordination [15, 53, 64]. Finally, Brown [7] studied prior OS and IS literature, finding a significant number of organizations engaged in horizontal mechanisms, i.e. a combination of centralized and non-centralized forms of authority command. These mechanisms facilitate collaboration and integration in the organization, and further promote the coordination of business and IS activities across corporate boundaries [7].

Formalization and standardization. Formalization and standardization is the extent to which policies and rules are documented and established through standard routines [49]. Formalization and standardization have been addressed by OS literature in the context of large organizations with complex work environments [36]. Within these complex environments, formalization and standardization act as a coordination mechanism through applying methods and procedures to reduce complexity in processes within, and collaborations between organizational units [36, 49]. Hereby, formalization and standardization lower costs/risks and increase the organization's overall efficiency [82]. OS studies see coordination to be achieved by standards in workflow processes [47, 55] as well as by the establishment of rules and formalized procedures [29, 38]. Both methods are reviewed as rather top-down delegation [27]. The same counts for behavioral control on interactions, which aims at fostering coordination across organizational units by rules and regulations [29, 58]. For this reason, strong hierarchical organizations are often described as being highly dependable on formalization, which establish structural methods top down, for example by the formation of routines [29, 38]. In line with formalization, this coordinating effect is similarly explained by standardization [27].

The IS discipline explicitly differentiates formalization/standardization from informal, more personal modes of coordinating corporate entities and stakeholders [7]. Compared to OS, the IS literature covers formalization and standardization in the context of IT support on complex organizational environments and workflow processes [48, 68]. Both formalization and standardization are addressed as approaches for organization-wide guidance/coordination, thereby incorporating standardized methods and technologies [48, 68] and formalizing the intra-organizational process alignment function of IT [72].

Planning. Systems and processes like strategic planning, budgeting, schedules, and goal settings are considered as planning mechanisms [49]. The OS literature investigates the coordination mechanism of planning in the case of new technology investment, team performance, and dynamics, as well as concurrent engineering processes. Kapoor and Lee [37] examine firms' coordination choices (alternative plans) and demonstrate alliance types of coordination (the broader the better) as the most effective plan for new technology investment. Lanaj et al. [43] criticize decentralized planning, owing to the fact that even though decentralized planning has positive effects on multi-team performance, it has even stronger negative impact on between-team dynamics. Finally, Terwiesch et al. [74] provide alternative coordination strategies to manage coordination in concurrent (parallel) engineering processes.

In terms of planning, the IS discipline is less prescriptive than OS. Except Tan and Harker [73], who explicitly recommend distributed scheduling methods, other studies aim at outlining a typology of planning mechanisms instead of prescribing a specific approach [16, 28]. Yet, other studies take a different perspective and outline steps of strategic planning for information resource management by taking into account coordination requirements [68] or argue the strategic opportunity granted by communication technology to foster coordination in a globally distributed teams [80].

Control. Control mechanisms consider both output control as well as bureaucratic and impersonal control [49]. The OS discipline provides different perspectives on control mechanisms of coordination. Considering control mechanisms as a feedback loop between different coordination practices [9], the existing literature suggests a taxonomy of control options (structures) for different sets of coordination practices [21, 71]. The extant literature also investigates control mechanisms on both the individual and group (team) levels. On the individual level, the role of liaisons [61] as well as the transfer of managers between different subsidiaries [18] has been discussed. On the group (team) level, scholars have investigated control mechanism as a means for knowledge sharing and integration [29, 62]. Lastly, organizational learning, as a control mechanism [10], has also been the focus of extant research.

In the IS literature, coordination and control are often discussed concurrently. A great deal of research has been dedicated to illustrate the role of IT as enabler of coordination and control [22, 59]. Also, a large number of publications discusses the trade-off between coordination and control [11, 22], where coordination is reflected in integrated/federated IS and control in centralized IS. Furthermore, in line with some topics of research in OS literature, IS scholars also provide a taxonomy of control mechanisms for business-to-IT and IT-to-IT units horizontal collaborations [7], discussing the organizational learning aspect of control mechanisms [69].

3.2 Informal Mechanisms

Lateral relations. Martinez and Jarillo [49] explain lateral relations as direct contact between individuals, groups or organizational departments that cut across the formal structure. OS literature sheds light on lateral relations by the investigation of task integration (activity-resource fit) and interactional behavior [7, 28]. Here, lateral relations are described as cross-departmental forms of interaction: complex tasks are coordinated by cross-functional interactions, thereby integrating tasks and resources [7, 9]. Another widely addressed topic refers to employee behavior and roles in lateral relations [55], indicating that they impact social interaction [76].

Regarding the IS discipline, lateral relations are mostly described in the context of task design, role-based interactions, and group dynamics with the purpose of conquering organizational/task complexity [2]. Thus, organizational structures and processes are of facilitating rather than impacting relations [28]: for instance, cross-departmental relations may be fostered by IS governance mechanisms [7, 16] and eventually lead to integrations of lateral interactions [44]. These lateral relations, which are considered as horizontal mechanisms [28, 70, 77], evolve as supplement mechanisms to the firm's established structural forces (i.e. hierarchy) and enable interaction across departments [4].

Communication. According to Martinez and Jarillo [49] and similar to lateral relations, communication acts as a supplement to formal mechanisms. Both OS and IS literature consider this coordination mechanism as contact practice among organizational actors that fosters information and knowledge exchange [57] as well as organizational learning [20]; however, the form and nature of underlying mechanisms differs. The OS literature investigates communication mechanisms as creator of a common basis to transfer information [55, 62] and knowledge [57]. Communication, in knowledge management, can also be an active coordinating mechanism in the practice of social interaction [76]. This coordination mechanism evolves in the form of boundary spanners [27], connecting social interactions and facilitating the exchange of information [38].

The IS literature mainly addresses the role of IT systems as a means to reinforce communication mechanisms [16], for instance, reducing informational complexity [65], and bridging differences in knowledge characteristics and physical distances [13, 16]. The role of IT has been highly emphasized in enhancing information processing capabilities, task, information, and environmental uncertainty [70]. Since information processing capabilities are highly dissimilar in nature [8], information technologies are considered as a necessity to realize communication mechanisms [13]. Furthermore, IT contributes to coordination performance through enabling and supporting cooperative work [70].

Socialization. Socialization is described as building an organizational culture of shared strategic objectives and values [49]. Socialization remains a subject to the OS discipline as a control artifact on employee behavior in the organizational environment [18, 31]. OS literature indicates socialization as coordination base in the context of interaction and relationship management [32]. The underlying mechanisms of socialization facilitate not only the coordination within organizations, but also increase firm capabilities, individual knowledge, and organizational learning [8, 29, 76].

The IS discipline introduces socialization as IT supported integration mechanism of the work environment [31]. Socialization is present in information processing and communication, both inside [35] and outside [17] of the organization. It furthermore coordinates organizational units toward shared objectives [3, 34]. Important to mention is the enablement and support of cooperative work through the means of IT [70]. In addition, socialization integrates work-flow processes through the support of IT and thereby reduces their complexity [19].

4 Discussion

Building on a comprehensive review of coordination in the extant OS and IS research, we synthesized the reflection of coordination mechanisms to three general streams (see Table 2).

The first stream describes *IS as an artifact subject to coordination*. Due to the substantial penetration of IS in organizations' daily routines [6] as well as in strategic planning processes [51], the integration of IS in organizations, from the coordination perspective, has long been the subject of investigations. The integration of large, dispersed organizations through the means of IT and IS artifacts is instrumental to realize coordination success, not only on a local/business unit basis, but also on the enterprise-wide level [16]. This finding is underpinned by a considerable number of studies in OS outlets [47, 74] that contribute to this discussion. In this stream, IS follows OS as its reference discipline in the main discourses, for instance in socialization, communication, and formalization/standardization [8, 31, 60].

The second stream exposes *IS as a means of coordination*. This is where IS act primarily as technological coordination support for organizational work environments — more prominently for communication [16] — such as by IT systems that bridge physical distances or different knowledge characteristics [13, 16]. In addition, IS also support the horizontal integration of the work environment, for instance by aligning corporate teams to a boundaryless network [34], by linking team members toward shared objectives [34], or by enabling cooperative work through the means of IT systems [70]. Nevertheless, IS literature goes beyond the mere discussion on IS as yet another coordination means. Due to ever increasing size of organizations and their presence in global markets, IS are used to leverage global synergies [22], to coordinate business functions [39], and to manage cross-subsidiary similarities [11]. This is where IS literature contributes to OS through commencing emergent management approaches, such as IS-enabled enterprise transformation, digital transformation, among the others. For instance, these new approaches are initially discussed in new typologies of control mechanisms [7], modular business configuration to overcome complexity [65], and IS-enabled horizontal integration [16].

The third stream reveals the *complementarity of coordination mechanisms in IS*. Typically, a high degree of specificity of coordination mechanisms helps to translate coordination goals into individual tasks and actions. However, this specificity might only be useful in a given situation [12] and it may neglect to consider the organization as a whole. For this reason, IS literature has often applied a combination of different perspectives of mechanisms, concluding that the reach and impact of specific/one-sided coordination mechanisms—for instance, strict top-down driven planning or control mechanisms—often remain limited [4, 7]. For example, formal coordination mechanisms are often complemented by informal mechanisms, in which desired coordination

Table 2. Main discourses on coordination mechanisms in OS and IS literature

	Mechanisms	OS Discipline	IS Discipline
Formal	Departmentalization	• Hierarchical organizational structure/design • Organizational process design	• Structural overlays and physical colocation as related to the IT/IS context • IT as a means to support hierarchies and departmentalization • IT support on task-related interactions and decreasing transaction costs
	Decision-Making	• Centralized decision-making: hierarchical design of organization • Decentralized decision-making: social interaction; organizational communication • Horizontal mechanisms: structural overlays and physical colocation	• IT support for integrating cross-unit decision-making processes • IT support to facilitate enterprise-wide decision-making in decentralized structures • Horizontal mechanisms: structural overlays and physical colocation
	Formalization/ Standardization	• Formalized standards, rules, routines, workflows, and policies • Formalization and standardization in complex organizational environment	• Formalization and standardization in complex organizational environment • IT support for workflows • Modular formalization • Standardization in technologies
	Planning	• Alternative coordination structures/strategies • Decentralized vs. centralized coordination planning	• Typology of coordination planning mechanisms • IT-supported planning methods
	Control	• Taxonomy of control structures • Individual and group (team) level control • Organizational learning	• Taxonomy of control mechanisms • IT as enabler of control • Organizational learning
Informal	Lateral Relations	• Organizational behavior • Cross-functional interaction • Activity-/Task-resource fit • Informal character • Physical colocation	• Horizontal interdependencies • Informal character • Governance/Integration spanner
	Communication	• Learning/knowledge exchange • Social interaction • Boundary spanners	• IT to reinforce communication mechanisms • IT to reduce complexity, facilitate knowledge sharing, and support cooperative work
	Socialization	• Culture, values, shared objectives • Interactional behavior • Firm capabilities and learning	• IT supported (internal and external) environmental integration • IT supported shared objectives

goals become concretized in specific tasks, actions, or sub-goals [4, 7, 16]. These findings highlight, on the one side, the complementarity of coordination mechanisms for prospective research. On the other side, these findings also imply to reconsider singular perspectives in research, which often shed light only on specific mechanisms, but neglect their meaning to the organization as a whole. For instance, enterprise-wide IS management disciplines, such as enterprise architecture management [1] or project portfolio management [14], might benefit from such a broader, complementary perspective.

5 Conclusion

This study offers a structured representation of coordination and its respective mechanisms in both OS and IS research. Through contrasting the reflection of coordination in the OS and IS discipline, this study argues how IS research follows and how it can go beyond its reference discipline of OS. Our investigation is limited to a selective, although important, set of peer-reviewed journals. We admit that the current evaluation neglects other potentially relevant contributions. Due to the chosen level of abstraction, more granular insights into the collected set of coordination mechanisms become a necessary step for future research. Nevertheless, this study provides a valuable basis given the large number of topically broad publications discussing coordination in various contexts.

With regards to the resulted insights from our review of coordination literature, we encourage future research to particularly focus on the second and third identified stream, namely, *IS as a means of coordination* and the (often necessary) *complementarity of coordination mechanisms. Firstly*, we emphasize to focus on decentralized and federated as well as informal modes of coordination. As mentioned earlier, companies are growing to target diverse geographical markets through diverse sets of products and services; simultaneously, they are impacted by a broad range of dynamic influences. Owing to the increasing investments in corporate IS, growing interdependencies as well as complexities, firmly centralized modes of coordination become unfeasible and need to be complemented. Decentralized (also horizontal) and federated forms are expected to complement and leverage coordination among sub-units and subsidiaries, thereby reinforcing cross-unit collaboration, synergies, and ultimately performance improvements. This might lead to eventually reconsider the traditional understanding of exercising coordination in a strict hierarchical, top-down driven way. This also implies the necessity of investigating how organizations dynamically move between different modes of coordination as well as how top-down and bottom-up modes of coordination co-exist or complement each other. Further, due to increasing dominance of technologies that foster informal coordination, such as enterprise social media, the impact of traditionally formal mechanisms might be reconsidered through the complementary impact of informal, more personal mechanisms. *Secondly*, we call for future research to deepening investigations into pertinent IS sub-disciplines that all share the same ends, i.e. to align local corporate IS endeavors so as to meet enterprise-wide objectives and long-term intentions. This encompasses, among others, the disciplines of enterprise engineering, IT governance, project portfolio management, and enterprise architecture management.

Acknowledgement. This work has been supported by the Swiss National Science Foundation (SNSF).

References

1. Aier, S., Labusch, N., Pähler, P.: Implementing architectural thinking. In: Persson, A., Stirna, J. (eds.) CAiSE 2015. LNBIP, vol. 215, pp. 389–400. Springer, Cham (2015). doi: 10.1007/978-3-319-19243-7_36
2. Andres, H.P., Zmud, R.W.: A contingency approach to software project coordination. J. Manag. Inf. Syst. **18**(3), 41–70 (2002)
3. Ba, S., Stallaert, J., Whinston, A.B.: Research commentary: introducing a third dimension in information systems design - the case for incentive alignment. Inf. Syst. Res. **12**(3), 225–239 (2001)
4. Balaji, S., Brown, C.V.: Lateral coordination mechanisms and the moderating role of arrangement characteristics in information systems development outsourcing. Inf. Syst. Res. **25**(4), 747–760 (2014)
5. Bordetsky, A., Mark, G.: Memory-based feedback controls to support groupware coordination. Inf. Syst. Res. **11**(4), 366–385 (2000)
6. Broadbent, M., Weill, P., Clair, D.S.: The implications of information technology infrastructure for business process redesign. MIS Q. **23**(2), 159–182 (1999)
7. Brown, C.V.: Horizontal mechanisms under differing is organization contexts. MIS Q. **23**(3), 421–454 (1999)
8. Brown, J.S., Duguid, P.: Knowledge and organization: a social-practice perspective. Organ. Sci. **12**(2), 198–213 (2001)
9. Bruns, H.C.: Working alone together: coordination in collaboration across domains of expertise. Acad. Manag. J. **58**(1), 62–83 (2013)
10. Ching, C., Holsapple, C.W., Whinston, A.B.: Reputation, learning and coordination in distributed decision-making contexts. Organ. Sci. **3**(2), 275–297 (1992)
11. Clemmons, S., Simon, S.J.: Control and coordination in global ERP configuration. Bus. Process Manag. J. **7**(3), 205–215 (2001)
12. Crowston, K.: A taxonomy of organizational dependencies and coordination mechanisms. In: Malone, T.W., Crowston, K., Herman, G.A. (eds.) Organizing Business Knowledge, pp. 85–108. MIT Press, Cambridge (2003)
13. Dabbish, L., Kraut, R.E.: Awareness displays and social motivation for coordinating communication. Inf. Syst. Res. **19**(2), 221–238 (2008)
14. De Reyck, B., Grushka-Cockayne, Y., Lockett, M., Calderini, S.R., Moura, M.: The impact of project portfolio management on information technology projects. Int. J. Project Manag. **23**(7), 524–537 (2005)
15. DeSanctis, G., Gallupe, R.B.: A foundation for the study of group decision support systems. Manag. Sci. **33**(5), 589–606 (1987)
16. DeSanctis, G., Jackson, B.M.: Coordination of information technology management: team-based structures and computer-based communication systems. J. Manag. Inf. Syst. **10**(4), 85–110 (1994)
17. Dibbern, J., Winkler, J., Heinzl, A.: Explaining variations in client extra costs between software projects offshored to India. MIS Q. **32**(2), 333–366 (2008)
18. Edström, A., Galbraith, J.R.: Transfer of managers as a coordination and control strategy in multinational organizations. Adm. Sci. Q. **22**(2), 248–263 (1977)
19. Espinosa, J.A., Slaughter, S.A., Kraut, R.E., Herbsleb, J.D.: Familiarity, complexity, and team performance geographically distributed software development. Organ. Sci. **18**(4), 613–630 (2007)
20. Faraj, S., Sproull, L.: Coordinating expertise in software development teams. Manag. Sci. **46**(12), 1554–1568 (2000)

21. Faraj, S., Xiao, Y.: Coordination in fast-response organizations. Manag. Sci. **58**(8), 1155–1169 (2006)
22. Finnegan, P., Longaigh, S.N.: Examining the effects of information technology on control and coordination relationships: an exploratory study in subsidiaries of pan-national corporations. J. Inf. Technol. **17**(3), 149–163 (2002)
23. Foss, N.J., Lyngsie, J., Zahra, S.A.: The role of external knowledge sources and organizational design in the process of opportunity exploitation. Strateg. Manag. J. **34**(12), 1453–1471 (2013)
24. Fritz, M.B.W., Narasimhan, S., Rhee, H.-S.: Communication and coordination in the virtual ofice. J. Manag. Inf. Syst. **14**(4), 7–28 (1998)
25. Galbraith, J.R.: Organization design: an information processing view. Interfaces **4**(3), 28–36 (1974)
26. Garicano, L., Wu, Y.: Knowledge, communication, and organizational capabilities. Organ. Sci. **23**(5), 1–16 (2012)
27. Gittel, J.H.: Coordinating mechanisms in care provider groups: relational coordination as a mediator and input uncertainty as a moderator of performance effects. Manag. Sci. **48**(11), 1408–1426 (2002)
28. Gosain, S., Lee, Z., Kim, Y.: The management of cross-functional inter-dependencies in ERP implementations: emergent coordination patterns. Eur. J. Inf. Syst. **14**(4), 371–387 (2005)
29. Grant, R.M.: Toward a knowledge-based theory of the firm. Strateg. Manag. J. **17**(Winter Special Issue), 109–122 (1996)
30. Harzing, A.W.K.: Journal Quality List. http://www.harzing.com/download/jql_subject.pdf. Accessed 1 February 2017
31. Horton, M., Biolsi, K.: Coordination challenges in a computer-supported meeting environment. J. Manag. Inf. Syst. **10**(3), 7–24 (1993)
32. Humphrey, S.E., Morgeson, F.P., Mannor, M.J.: Developing a theory of the strategic core of teams: a role composition model of team performance. J. Appl. Psychol. **94**(1), 48–61 (2009)
33. Jansen, J.P., van den Bosch, F.A.J., Volberda, H.W.: Managing potential and realized absorptive capacity: how do organizational antecedents matter? Acad. Manag. J. **48**(6), 999–1015 (2005)
34. Jarvenpaa, S.L., Knoll, K., Leidner, D.E.: Is anybody out there? Antecedents of trust in global virtual teams. J. Manag. Inf. Syst. **14**(4), 29–64 (1998)
35. Jarvenpaa, S.L.: Staples: exploring perceptions of organizational ownership of information and expertise. J. Manag. Inf. Syst. **18**(1), 151–183 (2001)
36. Joseph, J., Ocasio, W.: Architecture, attention, and adaptation in the multibusiness firm: general electric from 1951 to 2001. Strateg. Manag. J. **33**(6), 633–660 (2012)
37. Kapoor, R.: Coordinating and competing in ecosystems: how organizational forms shape new technology investments. Strateg. Manag. J. **34**(3), 274–296 (2013)
38. Kellogg, K.C., Orlikowski, W.J., Yates, J.: Life in the trading zone: structuring coordination across boundaries in postbureaucratic organizations. Organ. Sci. **17**(1), 22–44 (2006)
39. Kim, K., Park, J.H., Prescott, J.E.: The global integration of business functions: a study of multinational businesses in integrated global industries. J. Int. Bus. Stud. **34**(4), 327–344 (2003)
40. Kolde, E.J., Hill, R.E.: Conceptual and normative aspects of international management. Acad. Manag. J. **10**(2), 119–128 (1967)
41. Kretschmer, T., Puranam, P.: Integration through incentives within differentiated organizations. Organ. Sci. **19**(6), 860–875 (2008)
42. Lam, W.: Investigating success factors in enterprise application integration - a case-driven analysis. Eur. J. Inf. Syst. **14**(2), 175–187 (2005)

43. Lanaj, K., Hollenbeck, J.R., Ilgen, D.R., Barnes, C.M., Harmon, S.J.: The double-edged sword of decentralized planning in multiteam systems. Acad. Manag. J. **56**(3), 735–757 (2013)
44. Li, E.Y., Jiang, J.J., Klein, G.: The impact of organizational coordination and climate on marketing executives' satisfaction with information systems services. J. Assoc. Inf. Syst. **4**(1), 99–117 (2003)
45. Malone, T.W., Crowston, K.: What is coordination theory and how can it help design cooperative work systems? In: Proceedings of the Conference on Computer-Supported Cooperative Work, pp. 357–370 (1990)
46. Malone, T.W., Crowston, K.: The interdisciplinary study of coordination. ACM Comput. Surv. **26**(1), 87–119 (1994)
47. Malone, T.W., Crowston, K., Lee, J., Pentland, B.T., Dellarocas, C., Wyner, G.M., Quimby, J., Osborn, C.S., Bernstein, A., Herman, G.A., Klein, M., O'Donnell, E.: Tools for inventing organizations: toward a handbook of organizational processes. Manag. Sci. **45**(3), 425–443 (1999)
48. Mani, D., Srikanth, K., Bharadwaj, A.: Efficacy of R&D work in offshore captive centers: an empirical study of task characteristics, coordination mechanisms, and performance. Inf. Syst. Res. **25**(4), 846–864 (2014)
49. Martinez, J.I., Jarillo, J.C.: The evolution of research on coordination mechanisms in multinational corporations. J. Int. Bus. Stud. **20**(3), 489–514 (1989)
50. Massey, A.P., Montoya-Weiss, M.M., Hung, Y.-T.: Because time matters: temporal coordination in global virtual project teams. J. Manag. Inf. Syst. **19**(4), 129–155 (2003)
51. McAfee, A., Brynjolfsson, E.: Investing in the IT that makes a competitive difference. Harvard Business Review **86**(7/8), 98–107 (2008)
52. Melville, N., Kraemer, K., Gurbaxani, V.: Review: information technology and organizational performance: an integrative model of IT business value. MIS Q. **28**(2), 283–322 (2004)
53. Mentzas, G.N.: Coordination of joint tasks in organizational processes. J. Inf. Technol. **8**, 139 (1993)
54. Mintzberg, H.: The Structuring of Organizations: A Synthesis of the Research. Prentice-Hall, Englewood Cliffs (1979)
55. Mom, T.J.M., van den Bosch, F.A.J., Volberda, H.W.: Understanding variation in managers' ambidexterity: investigating direct and interaction effects of formal structural and personal coordination mechanisms. Organ. Sci. **20**(4), 812–828 (2009)
56. Murer, S., Bonati, B., Furrer, F.J.: Managed Evolution: A Strategy for Very Large Information Systems. Springer, Heidelberg (2010)
57. Nonaka, I.: A dynamic theory of organizational knowledge creation. Organ. Sci. J. Inst. Manag. Sci. **5**(1), 14–37 (1994)
58. Ouchi, W.G.: Markets, bureaucracies, and clans. Adm. Sci. Q. **25**(1), 129–141 (1980)
59. Ravichandran, T., Liu, Y., Han, S., Hasan, I.: Diversification and firm performance: exploring the moderating effects of information technology spending. J. Manag. Inf. Syst. **25**(4), 205–240 (2009)
60. Ren, Y., Kiesler, S., Fussell, S.R.: Interruptions, coping mechanisms, and technology recommendations. J. Manag. Inf. Syst. **25**(1), 105–130 (2008)
61. Reynolds, E.V., Johnson, J.D.: Liaison emergence: relating theoretical perspectives. Acad. Manag. Rev. **7**(4), 551–559 (1982)
62. Rico, R., Sánchez-Manzanares, M., Gil, F., Gibson, C.: Team implicit coordination processes: a team knowledge-based approach. Acad. Manag. Rev. **33**(1), 163–184 (2008)
63. Ring, P.S., van de Ven, A.H.: Developmental processes of cooperative interorganizational relationships. Acad. Manag. Rev. **19**(1), 90–118 (1994)

64. Roberts, N., Grover, V.: Leveraging information technology infrastructure to facilitate a firm's customer agility and competitive activity: an empirical investigation. J. Manag. Inf. Syst. **28**(4), 231–270 (2012)

65. Rosenkranz, C., Vraneši, H., Holten, R.: Boundary interactions and motors of change in requirements elicitation: a dynamic perspective on knowledge sharing. J. Assoc. Inf. Syst. **15**(6), 306–345 (2014)

66. Ross, J.W., Beath, C.M.: Sustainable IT outsourcing success: let enterprise architecture be your guide. MIS Q. Exec. **5**(4), 181–192 (2006)

67. Schmidt, C., Buxmann, P.: Outcomes and success factors of enterprise IT architecture management: empirical insight from the international financial services industry. Eur. J. Inf. Syst. **20**(2), 168–185 (2011)

68. Selig, G.J.: Approaches to strategic planning for information resource management (IRM) in multinational corporations. MIS Q. **6**(2), 33–45 (1982)

69. Sherif, K., Zmud, R.W., Browne, G.J.: Managing peer-to-peer conflicts in disruptive information technology innovations: the case of software reuse. MIS Q. **30**(2), 339–356 (2006)

70. Shih, H.-P.: Technology-push and communication-pull forces driving message-based coordination performance. J. Strateg. Inf. Syst. **15**(2), 105–123 (2006)

71. Sikora, R., Shaw, M.J.: A multiagent framework for the coordination and integration of information systems. Manag. Sci. **44**(11), 65–78 (1998)

72. Slaughter, S.A., Levine, L., Ramesh, B., Pries-Heje, J., Baskerville, R.: Aligning software processes with strategy. MIS Q. **30**(4), 891–918 (2006)

73. Tan, J.C., Harker, P.T.: Designing workflow coordination: centralized versus market-based mechanisms. Inf. Syst. Res. **10**(4), 328–342 (1999)

74. Terwiesch, C., Loch, C.H., De Meyer, A.: Exchanging preliminary information in concurrent engineering: alternative coordination strategies. Organ. Sci. **13**(4), 402–419 (2002)

75. Thompson, J.D.: Organizations in action: social science bases of administrative theory. McGraw-Hill, New York (1967)

76. Tsai, W.: Social structure of "Coopetition" within a multiunit organization: coordination, competition, and intraorganizational knowledge sharing. Organ. Sci. **13**(2), 179–190 (2002)

77. Van de Ven, A.H., Delbecq, A.L., Koenig Jr., R.: Determinants of coordination modes within organizations. Am. Sociol. Rev. **41**(2), 322–338 (1976)

78. Webster, J., Watson, R.T.: Analyzing the past to prepare for the future: writing a literature review. MIS Q. **26**(2), 13–23 (2002)

79. Weigelt, C., Miller, D.J.: Implications of internal organization structure for firm boundaries. Strateg. Manag. J. **34**(12), 1411–1434 (2013)

80. Wiredu, G.O.: Understanding the functions of teleconferences for coordinating global software development projects. Inf. Syst. J. **21**(2), 175–194 (2011)

81. Xue, L., Ray, G., Gu, B.: Environmental uncertainty and IT infrastructure governance: a curvilinear relationship. Inf. Syst. Res. **22**(2), 389–399 (2011)

82. Zhou, Y.M.: Synergy, coordination costs, and diversification choices. Strateg. Manag. J. **32**(6), 624–639 (2011)

Author Index